SCIENCE COMMUNICATION IN SOUTH AFRICA

Reflections on Current Issues

Edited by
Peter Weingart, Marina Joubert & Bankole Falade

Published in 2019 by African Minds
4 Eccleston Place, Somerset West 7130, Cape Town, South Africa
info@africanminds.org.za
www.africanminds.org.za

Funding acknowledgement: This work is based on the research supported by the South African Research Chairs Initiative of the Department of Science and Technology and National Research Foundation of South Africa (grant number 93097). Any opinion, finding and conclusion or recommendation expressed in this material is that of the authors and the NRF does not accept any liability in this regard.

Contents

Acknowledgement

This book was made possible by and was published under the auspices of the South African Research Chair in Science Communication hosted by the Centre for Research on Evaluation, Science and Technology (CREST), Stellenbosch University. We thank the Department of Science and Technology (DST) and the National Research Foundation (NRF) for their generous support.

1 Introduction

Peter Weingart, Marina Joubert & Bankole Falade

Why science communication?

To understand the surge of activities nowadays termed 'science communication' one has to get a sense of the volume and speed of development of science over the last century, as well as its place in society. Science, whether measured in terms of scientists or in terms of scientific publications, has grown exponentially since the birth of modern science in the late 17th century. While this dynamic growth went unnoticed for a while, starting out from just a few adherents to the new ways of gaining knowledge, it became a subject of systematic reflection only in the middle of the 20th century when the US historian of science and father of bibliometrics famously noted that 90% of all scientists that had ever lived were alive at present (Price, 1963). Even though scientists (and engineers) had already contributed considerably to economic development during the late 19th century, their numbers and their impact on societies really began to matter politically and economically during and after the First World War. It was not until after the Second World War that science policy became a separate field of policy-making, first in the US, and then in Europe, Japan and Australia. Until then science was an activity carried out in relative isolation from the rest of society. Scientists communicated among themselves, within their disciplinary communities and in languages that became more and more opaque as their fields became ever more specialised. Science was, in the words of Don K. Price, *exceptional* in the sense that it was the only institution that received public funds without having to account for it (Price, 1965).

This exceptionalism began to fade already in the mid-1950s when large technology projects – civilian nuclear power, aerospace and data processing – highlighted the economic utility of science (and technology). The first attempts at improving the 'public understanding of science' in the US were motivated primarily by concerns related to the Cold War: an apparent lack of STEM students threatening the effort of the country to prevail in the competition for technological leadership and the need to secure public support for the space programme. These two motives of science policy have become generalised beyond the original context, they underlie science communication policies in virtually all countries that have such policies, and they are present to this day to contribute to innovation and to secure public acceptance of public expenditures for science, as well as the implementation of new technologies.

Public acceptance of expenditures for scientific research was particularly critical. The then dominant so-called 'linear model' of innovation stipulated that all economic innovation emerged from prior basic research, that the direction of such research was to be determined by scientists only, and that the outcome of research could not be predicted (Bush, 1945). This constellation was at the heart of the exceptionalism of science, and it was supported by the political context in which the freedom of science was to symbolise the superiority of the West. The need to secure public consent became more urgent as science budgets grew to politically visible dimensions, ultimately reaching 2–3% of GDP in the wealthier countries.[1] Consequently, the general public, in the form of the electorate, had to be addressed to trust the scientific community's decisions and to legitimate R&D expenditures.

The general public had been addressed by scientists long before.

---

1 Cf. https://data.oecd.org/rd/gross-domestic-spending-on-r-d.htm

In fact, at the inception of modern science, scientists attempted to capture the interest and fascination of the aristocracy on which they depended for support. During the second half of the 19th century the popularisation of science almost became a separate profession. Alexander von Humboldt, addressing the educated bourgeoisie and the working class in his Kosmos Lectures, turned out to be the instigator of the first 'science centre' *avant la lettre*, the Urania in Berlin. The spirit of popularisation that was very much also a spirit of enlightenment which could thrive as long as the science of the day was 'accessible' to the lay public, at least in principle, a condition that eroded with the increasing abstractness of concepts, language and subject matters in many fields heralded by quantum mechanics at the beginning of the 20th century (Bensaude-Vincent, 2001).

The advent of 'public understanding of science' was thus characterised by a mix of motives: economic, political, legitimation and enlightenment of the public. Over the years many actors have joined in efforts to improve the public's understanding of science, but to this day there is no consensus among scholars about its goals, or about the criteria of success or failure (Lewenstein, 2003: 1). In 1985, the British Royal Society published its so-called *Bodmer Report* which urged the 'Economic and Social Research Council (ESRC) and other appropriate bodies to devise methods of monitoring attitudes to science in the United Kingdom' (Bodmer, 1985: 31). More than three decades since its publication there are still no adequate methods in place, nor are effective steps undertaken, to evaluate the many activities that are carried out under the label of science communication (Short, 2013: 40).

This state of affairs is reflected in a lively debate published in a number of scholarly journals founded since the late 1970s (*Science Communication* in 1979; *Public Understanding of Science* in 1992; *Journal of Science Communication [JCOM]* in 2002). The original concerns among scientists focused on the knowledge among the public of basic scientific concepts, theories and methods. Surveys designed to gauge that knowledge (first by the US National Science Foundation) found that the public's understanding of science (PUS) – as defined by them – was extremely limited. It was believed that by identifying such 'deficits' of scientific 'literacy' the respective educational programmes could remedy this state and, as a result, by improving the public's understanding, this would also result in generally positive attitudes toward science. The so-called 'deficit-model' underlying the PUS approach was subsequently criticised both for its simplistic assumptions of information processing, but also for its paternalistic outlook on the relationship between science and the public. As time went by, the academic discussion of the right and effective format has moved from the deficit model to the 'contextual', the 'lay expertise' and finally the 'public participation models' (Lewenstein, 2003). The hitherto latest development in this evolution of science communication is the concept of 'public engagement with science and technology' (PEST) which propagates 'dialogical' formats between science and the public, active participation of citizens in science policy decision-making and even in research projects (flagged as 'citizen science') (Smallman, 2018; Stilgoe et al., 2014). Thus, the trajectory from 'deficit' to 'dialogue' appears to be one of greater proximity of science to the public, of 'inclusion' if not of a democratised relationship. However, the reality on the ground looks much more modest, and the lacuna between it and the lofty rhetoric of science policy programmes and idealistic scientists is the rationale for a 'science of science communication' (NAS, 2017).

Some challenges to science communication

Science communication programmes have become part of science policy for more than three decades in virtually all developed and in some developing countries, such as South Africa. Yet, in spite of the considerable cost incurred, there is still no serious evaluation of their effectiveness. Surveys of trust in science have remained methodologically weak and are rarely linked directly to particular communication programmes. This abstinence can be explained by at least two factors. First, there are a multitude of different motives that drive science communication programmes and that do not allow for the definition of distinct criteria against which to measure effects. Second are the vested interests of the actors that initiate the larger share of the programmes for purposes of public relations (Weingart & Joubert, 2019). In particular, universities are a pertinent example since they are organisations that compete for public funds, students and, most importantly, positive general attention that is expected to enhance their legitimacy with the public and policy-makers. Thus, their communication activities that used to be focused on (and limited to) press releases about new discoveries in their research laboratories have more often than not developed into public relations type communication, reflected in a dramatic growth of communication professionals at universities. Consequently, the focus of research institutions, universities and individual researchers is increasingly shifting from information/knowledge transfer to reputation control and image building (ALLEA, 2019; Schäfer, 2017). The resulting problem is that 'interested communication' commands considerably less trust (Heyl, 2018; Peters, 2015; Weingart & Guenther, 2016).

Another challenge to science communication that also impacts the public's trust is the role of social media. Social media platforms, above all Facebook and Twitter, have facilitated direct communication without the traditional journalistic intermediaries and have been greeted by universities and scientists alike for their promise to expand their reach and to capture public attention at a dimension that was unthinkable before their advent. The initial enthusiasm that the platforms have triggered has waned somewhat in view of the various scandals of data abuse for political and economic purposes, as well as – perhaps even more importantly – because of the spread of false information. The activities of anti-vaccination groups online have demonstrated the downside of the technology in an area of science communication which is particularly vulnerable, namely health communication, because it affects the medical well-being of individuals and entire communities (see Van Schalkwyk in this volume). Thus, the many positive and negative implications of the internet and social media for science communication have already attracted much research attention and will continue to do so in the foreseeable future (ALLEA, 2019; Van Dijck, Poell & De Waal, 2018).

The ultimate challenge to science communication results from a fundamental structural problem. None of the motives that drive the various science communication programmes provide clear-cut criteria that inform the selection of the content of what is to be communicated to the general public. Only particular stakeholder groups have clearly defined interests in what of the almost endless stock of accumulated scientific knowledge is relevant to them. It is to be expected, therefore, that such groups will form as clienteles for science communication, even where they do not already exist. (It is a common experience of museums, general science exhibitions and even more so of citizen science projects that they preach to the converted.) Unless all of science communication is transformed into a giant 'edutainment' project in which entertainment plays a dominant role and education a minor one, it is anything but clear how this problem may be solved. It could well be that the lofty engagement programmes that purportedly address the entire population will have to face the reality of reaching only those segments that are already engaged to some extent. For others, choices may have to be made about the information to be communicated to them on the basis of their everyday lives, the socio-economic contexts in which they live, and their immediate needs and interests (Guenther et al., 2018). This is particularly relevant in a country such as South Africa, where large parts of the population do not have the luxury, nor the educational background, to enjoy demonstrations of pure science without regard of its benefits to them.

For these very reasons we may witness (and even propagate) that the science communication programmes, above all in the developing countries, revert to a kind of engagement model with an inbuilt pragmatic focus. The design of these programmes, i.e. of their objectives and their contents, should be based on prior research into the needs, perceptions and expectations of different segments of the population. In this way, the interests of these groups are taken seriously, and they can be given a voice in various ways to assure that they are not being misinterpreted or otherwise distorted. This will make science communication more relevant to them, and it will prevent the communication of science from eroding into an exercise in the self-praise of science. How, then, has South African science policy taken up science communication?

Science communication in the South African context

Within two years of coming to power in 1994, South Africa's first democratic government adopted a Science and Technology White Paper that emphasised the need for a society which understands and values science as a facilitator of socio-economic progress (DACST, 1996). The new government wanted its citizens to be able to monitor policy, learn, collaborate, campaign and react to proposed legislation. However, given the political history of the country, where science remained isolated from the majority of South Africans, achieving a scientifically literate society presented a momentous challenge (Du Plessis, 2017).

The 'Year of Science and Technology' in 1998 was the first major science awareness campaign with the broad aim of 'demystifying' science through exhibitions, science shows and public talks in each of the nine provinces of South Africa. The government urged scientists and research organisations to help advance public awareness and appreciation of science. In 2002, the South African Agency for Science and Technology Advancement (SAASTA) was established to coordinate national science engagement activities. SAASTA took charge of science weeks, as well as a suite of science competitions and topic-specific awareness programmes, while also managing government support for science festivals and a network of science centres.

Subsequent policies highlighted public understanding of science as a prerequisite for South Africa to become a more innovative society with a more democratic and participatory mode of science governance than that which had been the norm throughout its history (DST, 2007). In 2015, the Department of

Science and Technology (DST) adopted a new science engagement framework designed to coordinate an ambitious portfolio of activities across all government departments, higher education institutions, science councils, museums and private sector partners (DST, 2015). This strategy positions science engagement as something that will enrich and improve people's lives, and seeks to develop a society that is knowledgeable about science, is scientifically literate and capable of forming opinions about science issues (DST, 2015). The government's commitment to public science engagement is also highlighted in the 2019 White Paper on Science, Technology and Innovation (DST, 2019). This policy spells out a number of ways in which future science engagement activities will be mandated and coordinated. Moreover, it accentuates the need for specialised training to develop the engagement and communication skills of journalists, scientists, students, learners, educators and science interpreters.

As is the case in many developing countries, efforts to promote a culture and understanding of science in South Africa face a number of significant societal challenges. South Africa is one of the most unequal societies in the world, with up to half of its nearly 60 million citizens living in chronic poverty.[2] Economic growth is tardy[3] and weighed down by the destructive effects of corruption, as well as infrastructure challenges and poor service delivery, in particular the ongoing risk of power outages. The official unemployment rate in the country hovers around 28%.[4] This situation necessitates the government to balance investment in science and education with pressing societal needs for housing, social security and healthcare.

---

2 With a Gini coefficient of 0.63 in 2015, the report describes South Africa as the most unequal country on earth; see https://www.iol.co.za/news/south-africa/south-africa-worlds-most-unequal-society-report-14125145

3 The South African economy grew by 0.8% in 2018, see http://www.statssa.gov.za/?p=11969

4 See https://tradingeconomics.com/south-africa/unemployment-rate

When it comes to the education system of the country, the picture is equally grim. The dire state of literacy amongst school learners in South Africa was revealed in the 2016 Progress in International Reading Literacy Study (Mullis et al., 2016) which ranked South Africa last out of 50 countries for its level of child literacy. Similarly, the generally poor performance of South African learners in mathematics and science is revealed in the 2015 Trends in International Mathematics and Science Study (Mullis et al., 2015). The mathematics performance of South African Grade 4 learners was rated 49th out of the 50 participating countries.

The 2018 'State of the Newsroom' (Finlay, 2018) report also reveals reasons for concern regarding the future ability of South African journalists to make a meaningful contribution to the public communication of science in the country. Traditional newsrooms are weakening and newspaper circulation is continuing a downward spiral, with some online media business models also failing. More and more journalists, including some experienced science journalists, are being retrenched from their jobs and forced into a so-called 'gig' economy in order to make a living.

The immense socio-economic and infrastructural challenges outlined above mean that it will be a daunting task for the South African government to achieve its ambitious goals in terms of public science engagement. The immensity of these challenges is further evident from recent research about the relationship between science and the South African society, as outlined above.

Science communication research in South Africa

Since the early 1990s, some South African researchers at the National Research Foundation (NRF) and the Human Sciences Research Council (HSRC) initiated a few small studies on scientific literacy, public understanding of science and public attitudes to science (see, for example, Blankley and Arnold, 2001; Pouris, 1991, 1993, 2003). More recently, the HSRC commissioned a number of larger surveys on public perceptions of science (Reddy et al., 2013), astronomy (Roberts et al., 2014) and biotechnology

(Gastrow et al., 2018). Investigations into representations of science in the media focused on biotechnology (Gastrow, 2010) and the Square Kilometre Array radio telescope (Gastrow, 2015, 2017). This body of research has highlighted the diversity of views and complex blend of perceived benefits and risks that South Africans associate with science (Guenther & Weingart, 2016; Reddy et al., 2013), while the influence of culture and the cultural distance to science is also evident (Guenther & Weingart, 2018; Guenther et al., 2018). Overall, a bleak picture of low interest in science and low involvement in public science activities emerges (Parker, 2017), but it is also clear that the unique history and challenges of the country motivate and encourage some scientists to reach out and engage with the public (Joubert, 2018).

The establishment of two South African research chairs in the field of science communication in 2015 provided the first foothold for research and academic training in this field in a university environment. Stellenbosch University hosts the South African Research Chair in Science Communication, while the South African Research Chair in Biotechnology Innovation and Engagement was established at Rhodes University. These two research chairs focus on different areas of the science–society interface. At Stellenbosch, research focuses on public perceptions and expectations of science, science communication via mass and social media, and institutional science communication. At Rhodes, the focus is on models of science engagement between scientists and the public, and the benefits of science engagement for researchers and communities.

Inter-disciplinary research focused on inclusivity, on transformation and on policy in the field of science–society engagement is also ongoing at the HSRC, in particular in its Social Policy, Knowledge Mobilisation and Impact Assessment Research Programme.

With its emphasis on science communication as an integral part of its science policy, South Africa has followed many other nations that are more advanced economically and less polarised in their social structures. While 'imitating' science policy institutions is a well-known effect that has been observed in many countries across the globe, it is both courageous and risky. It is courageous because it propels the country into modernity and sets landmarks of future development that may guide the work for generations to come. This pertains, in particular, to the early emphasis on the role of science and scientific literacy for a democratic society and for informed decision-making by its citizens which is more pronounced than in most other countries. However, it is risky at the same time as the obstacles to realising this ideal state of affairs are gargantuan, and persistent failure to bridge the gap between the idealistic rhetoric and reality may lead to frustration and cynicism. This is the specific challenge science communication research faces in South Africa.

In November 2018, the South African Research Chair in Science Communication at Stellenbosch University hosted a conference which was designed to focus on the challenges and prospects of science communication in a democratic South Africa. In particular, we wanted to reflect on the state of science communication research as a newly emerging field of scholarship in the country. The conference provided a platform for local researchers and invited practitioners of science communication to present their work and ideas, while the participation of global leaders in the field allowed valuable opportunities for exchange of information on new developments, networking and capacity-building.

Rather than publishing a collection of conference presentations, the articles assembled in this volume are a selection of those that addressed what we considered the most pressing issues of science communication in South Africa giving special attention to views and experiences from practitioners who are faced with problems 'on the ground'. The contributors were asked to re-write their presentations in any way they deemed appropriate to fit the format and focus of this book.

The chapters

Janice Limson's chapter argues that modern day challenges in science, engineering and technology call for new models of engagement between scientists and the public. These new models, which offer a more active role for the public in the process of scientific research, are at the centre of the European Union's Responsible Research and Innovation (RRI) framework. The chapter describes approaches for direct engagement of the public in shaping research at a university and uses biotechnology as a case study to explore the concepts of co-creation, participatory research and citizen science.

Models of public engagement with biotechnology explored include: direct communication between scientists and the public at a science fair; public involvement in laboratory based research; user surveys to elicit public views about new products; and engaging specific publics regarding their perspectives on current and future research.

Penelope S. Haworth and *Anne M. Dijkstra* continue to explore how science engagement and communication can contribute to putting RRI into practice in South Africa using findings from a European-funded project titled 'New Understanding of Communication, Learning, and Engagement in Universities and Scientific Institutions (NUCLEUS)' which ran from 2015 to 2019. The project found that while enthusiasm for engagement was high at the individual level, there were constraints imposed by budget, diversity and access to education. At the governmental level, where innovation is seen as a driver of economic advancement and living standards, equal access and inclusion are seen as challenges to the science system, but there are programmes which show benefits of the collaboration between science and society. At the institutional level, the project found that, despite commendable efforts, focus in fostering science education and outreach programmes, rural populations remain hard to reach and impact and engagement are not yet part of key performance indicators. Also, while research organisations embrace open access policies, and impact and engagement are considered important, they are not yet part of the key performance indicators for researchers.

The chapter by *Konosoang Sobane* and *Wilfred Lunga* focuses on behavioural change strategies for culturally diverse communities, arguing for the use of theories rooted in social psychology in the development of health communication approaches. Distinguishing between 'culturally sensitive' approaches which focus more on adaptation and 'culture-centred' approaches which harness culture-specific knowledge of the target communities and employ co-creation and co-development of communication strategies, the authors argue for a combination of both approaches for better results. They propose a framework for 'inclusion' which includes conducting a needs assessment, and involving segments of the population in the development of strategies to ensure cultural appropriateness. This means using communication platforms, which already have wide reach in the community, and disseminating information through locals who speak local languages and can respond to questions in culturally appropriate ways.

Doug S. Butterworth examines a special case of science communication, namely expert advice to policy-makers in a specific field: fisheries management. The chapter outlines the process of developing scientific advice and of its transmission to decision-makers in South Africa and internationally. The chapter argues that securing good outcomes depends on the efficiency of communicating findings of scientific analyses through the various stages of the process.

The chapter identifies poor science communication skills of scientists to both laypersons and decision-makers as a major problem. It argues that scientists need to broaden their presentation skills to include other stakeholders outside the scientific community who are unfamiliar with the scientific method. It also calls for an increase in the interaction between scientists on the one side, and stakeholders and decision-makers on the other, in 'intermediary groups' within the fisheries management decision structures to forge better understanding.

Shirona Patel's chapter deals with the role of social media in science communication. In the 'post truth' and 'fake news' environment where the traditional media is facing declining audiences, she explores how scientists are using social media; how they create compelling social media content; what the benefits are of using social media; and what the barriers and risks are. She goes on to discuss how, given the prevalence of the traditional media and its unique attributes, scientists develop strategies to combine both forms.

These issues are pertinent given that digitisation has already transformed newsrooms and the way in which science is communicated. Further changes to science communication practices should be expected with advances in Informatics and other Artificial Intelligence programmes. The emerging technologies, the chapter argues, provide endless opportunities to develop new creative approaches using multimedia technologies across multiple platforms, in real time and across physical and virtual boundaries. However, they come with associated risks, some of which are already known, and others which we can only predict.

George Claassen's chapter also deals with the impact of social media on science communication. He notes that the social media has further complicated efforts at making the public understand evidence-based science and there is an urgent need to separate it from pseudoscience or non-science, particularly in the field of health. The phenomenon, he writes, has become a growing concern not only to scientists but also to journalists and the society at large as fraudulent messages often go viral on the social media with grave consequences for health and well-being.

The problem is compounded when celebrities, with their wide following, spread false health messages in Twitter. Scientists too are increasingly using Twitter to communicate their works and the Ebola outbreak in West Africa confirms that the platform can be used both for accurate scientific information and for misinformation. The anti-vaccination debate is another case in point. Claassen concludes that there can be immense benefit for society in general if informed journalists and scientists engage with the public on Twitter to point out the harm quackery and pseudoscientific assertions can cause.

François van Schalkwyk deals especially with the risks posed to science communication when using social media. He examines the use of social media by the anti-vaccination movement in the context of communication networks, trust, open science and the norms of science. The research sought to create an empirically-based understanding of a fast-changing digital world which has increased access by non-scientists to the formal communication of science.

The research found that a highly active minority group (the anti-vaccination movement) uses selected scientific information to produce and amplify uncertainty in the broader population using social media. The chapter argues that the social media environment, which is devoid of scientific norms to steer action toward the establishment of truth, provides an ideal communication substrate, as does the networked nature of online communications. The chapter calls for more research into the potential risks and benefits of open science in the social media communication environment with a view to more generalisable insights.

Eric Allen Jensen addresses the issue of the evaluation of science communication. What counts as effective science communication? What difference is science communication making? How can it be measured whether the communication approach was effective at developing impact? These questions answer the overarching issue for science communication activities: impact.

Jensen argues that there is currently a lack of consensus on what counts as successful impact. He argues that a lack of good evaluation practices, poor survey design, inadequate training of science communicators and clarity of objectives are some of the main obstacles to an effective evaluation process. He calls for greater commitment to an improvement in survey design and the acceptance by practitioners that evaluation efforts should start from a neutral standpoint and be open to both positive and negative outcomes.

References

ALLEA (All European Academies) (2019). *Trust in Science and Changing Landscapes of Communication*. Berlin.

Bensaude-Vincent, B. (2001). A genealogy of the increasing gap between science and the public. *Public Understanding of Science,* 10, 99–113.

Blankley, W. & Arnold, R. (2001). Public understanding of science in South Africa: Aiming for better intervention strategies. *South African Journal of Science,* 97(3/4), 65–69.

Bodmer, W. F. (1985). *The Public Understanding of Science.* Report of a Royal Society ad hoc Group endorsed by the Council of the Royal Society. https://royalsociety.org/-/media/Royal_Society_Content/policy/publications/1985/10700.pdf.

Bush, V. (1945). *Science: The endless frontier.* Washington DC: Government Printing Office.

DACST (Department of Arts, Culture, Science and Technology) (1996). *White Paper on Science & Technology.* http://www0.sun.ac.za/scicom/wp-content/uploads/2019/04/white_paper_st_1996.pdf.

DST (Department of Science and Technology) (2007). *Innovation Towards a Knowledge-Based Economy: Ten-year plan for South Africa (2008–2018).* Pretoria. http://unpan1.un.org/intradoc/groups/public/documents/CPSI/UNPAN027810.pdf.

DST (Department of Science and Technology) (2015). *South African National Survey of Research and Experimental Development. Main analysis report 2014/15.* http://www.hsrc.ac.za/en/research-outputs/view/8613.

Du Plessis, H. (2017). Politics of science communication in South Africa. *Journal of Science Communication,* 16(3), A03. doi: 10.22323/2.16030203.

Finlay, A. (2018). *The State of the Newsroom Report.* Johannesburg: Wits Journalism.

Gastrow, M. (2010). *The Public Understanding of Biotechnology in the Media: A report for the National Advisory Council on Innovation and the National Biotechnology Advisory Committee.* Pretoria: Human Sciences Research Council.

Gastrow, M. (2015). Science and the social media in an African context: The case of the Square Kilometre Array telescope. *Science Communication,* 37(6), 703–722. doi: 10.1177/1075547015608637.

Gastrow, M. (2017). *The Stars in Our Eyes: Representations of the Square Kilometre Array Telescope in the South African Media.* Pretoria: HSRC Press.

Gastrow, M., Roberts, B., Reddy, V. & Ismail, S. (2018). Public perceptions of biotechnology in South Africa. *South African Journal of Science,* 114(1/2):Art. #2017–0276. doi: 10.17159/sajs.2018/20170276.

Guenther, L. & Weingart, P. (2016). A unique fingerprint? Factors influencing attitudes towards science and technology in South Africa. *South African Journal of Science,* 112(7), 8–11.

Guenther, L. & Weingart, P. (2018). Promises and reservations towards science and technology among South African publics: A culture-sensitive approach. *Public Understanding of Science,* 27(1), 47–58.

Guenther, L., Weingart, P. & Meyer, C. (2018). 'Science is everywhere, but no one knows it': Assessing the cultural distance to science of rural South African publics. *Environmental Communication,* 12(8), 1046–1061.

Heyl, A. (2018). Science communication vs. public relations: The potential effect of university press releases and the changing media landscape on science journalism in South Africa. Unpublished MPhil thesis, Stellenbosch University.

Joubert, M. (2018). Country-specific factors that compel South African scientists to engage with public audiences. *Journal of Science Communication,* 17(4), C04.

Lewenstein, B. (2003). Models of public communication of science and technology. https://ecommons.cornell.edu/bitstream/handle/1813/58743/Lewenstein.2003.Models_of_communication.CC%20version%20for%20Cornell%20eCommons.pdf?sequence=3&isAllowed=y.

Mullis, I. V. S., Martin, M. O., Foy, P. & Hooper, M. (2015). *Trends in International Mathematics and Science Study (TIMMS).* Boston, MA: TIMSS & PIRLS International Study Center.

National Academies of Sciences, Engineering and Medicine (NAS) (2017). *Communicating Science Effectively: A research agenda.* Washington, DC: The National Academies Press. doi: 10.17226/23674.

Parker, S. (2017). Development of indicators for the measurement of the South African public's relationship with science. Unpublished PhD thesis, Stellenbosch University.

Peters, H. P. (2015). *Science Dilemma: Between public trust and social relevance.* EuroScientist. http://www.euroscientist.com/trust-in-science-as-compared-to-trust-in-economics-and-politics.

Pouris, A. (1991). Understanding and appreciation of science by the public in South Africa. *South African Journal of Science,* 87(7), 358–359.

Pouris, A. (1993). Understanding and appreciation of science among South African teenagers. *South African Journal of Science,* 89(2), 68–69.

Pouris, A. (2003). Assessing public support for biotechnology in South Africa. *South African Journal of Science,* 99(11/12), 513–516.

Price, D. J. d. S. (1963). *Little Science, Big Science.* New York: Columbia University Press.

Price, D. K. (1965). *The Scientific Estate.* Cambridge: Belknap Press.

Reddy, V., Gastrow, M., Juan, A. & Roberts, B. (2013). Public attitudes to science in South Africa. *South African Journal of Science,* 109(1), 1–8. doi: 10.1590/sajs.2013/1200

Roberts, B., Struwig, J., Ngungu, M. & Gordon, S. (2014). Attitudes towards astronomy and the Square Kilometre Array (SKA) in South Africa. Tabulation report based on the 2013 round of the South African Social Attitudes Survey (SASAS) prepared for the Education and Skills Development (ESD) Research Programme. Pretoria: Human Sciences Research Council.

Schäfer, M. S. (2017). How changing media structures are affecting science news coverage. In K. Hall Jamieson, D. Kahan & D. Scheufele (eds), *Oxford Handbook on the Science of Science Communication* (pp. 51–60). New York: Oxford University Press.

Short, D. B. (2013). The public understanding of science: 30 years of the Bodmer Report. *The School Science Review,* 95(350), 39–44.

Smallman, M. (2018). Science to the rescue or contingent progress? Comparing 10 years of public, expert and policy discourses on new and emerging science and technology in the United Kingdom. *Public Understanding of Science,* 27(6), 655–673. doi: 10.1177/0963662517706452.

Stilgoe, J., Lock, S. J. & Wilsdon, J. (2014). Why should we promote public engagement with science? *Public Understanding of Science,* 23(1), 4–15. doi: 10.1177/0963662513518154.

Van Dijck, J., Poel, T. & De Waal, M. (2018). *The Platform Society: Public values in a connective world.* New York, NY: Oxford University Press.

Weingart, P. & Guenther, L. (2016). Science communication and the issue of trust. *Journal of Science Communication,* 15(5), C01.

Weingart, P. & Joubert, M. (2019). The conflation of motives of science communication: Causes, consequences, remedies. *Journal of Science Communication,* 18(3), Y01.

2 Engaging the public in scientific research: Models, prospects and challenges from the perspective of scientists

Janice Limson

Overview

Climate change, drought and desertification, crop failures, drug-resistant bacteria, invasive species, maternal and foetal mortality rates – the list goes on. Science, engineering and technology carries the hopes of a generation faced with a litany of grand challenges. In meeting those challenges, a 'new contract between science and society which encourages greater connectivity between the academic community and the rest of society' (Tassone et al., 2017: 338) is needed. This changing paradigm calls for new models and approaches in the training of scientists within universities.

In traditional modes of engagement between scientists and the public, the role of the public has largely been that of a passive recipient of scientific research, technological products and knowledge. Such deficit models have made way for more direct and engaged forms of communication between scientists and the public.

A growing school of thought extends this scientist-public dialogue further, advocating for the general public to assume a more active role in the process of scientific research itself, noting the potential that this may hold for enhancing the science, technology and engineering landscape. This thinking is at the centre of the European Union's Responsible Research and Innovation

(RRI) framework (European Commission, 2019), which calls for direct involvement of the public such that research is responsive to society, conducted not just in society but, more importantly, with and for society (Owen et al., 2012).

One of the challenges faced by concepts and notions of engaging the public in research is its 'in principle' adoption and uptake by scientists. Considering that the greatest proportion of scientific research takes place in universities, a specific challenge is the integration of direct public engagement into existing and future research, innovation and teaching programmes at universities. Scant research has explored the practical implementation of RRI and what these concepts mean in practice for both scientists and the public (Ribeiro et al., 2017).

Viewed through the perspective of research in universities in South Africa, this chapter describes approaches for direct engagement of the public in shaping research in a higher education institution using biotechnology as a case study. The study also explores in brief concepts of co-creation, participatory research and citizen science as models and tools to support RRI.

Responsible Research and Innovation (RRI)

The European Union's RRI framework advocates for involving the public in research and innovation, preferably at the earliest phases of the research cycle. Several definitions allude to the anticipated outcomes thereof with respect to sustainable research and innovation processes resulting in outcomes which have not only direct societal benefit but lead to successful and marketable products emanating from the innovation.

The European Commission references the need for adopting RRI principles in scientific work, such that these are not just inclusive, but sustainable: 'responsible research and innovation is an approach that anticipates and assesses potential implications and societal expectations with regard to research and innovation, with the aim to foster the design of inclusive and sustainable research and innovation' (European Commission, 2019: n.p.).

Von Schomberg's definition of RRI (2012: 9) references core values of ethics and processes that enhance the value of innovation itself, and its products: '[RRI] is a transparent, interactive process by which societal actors and innovators become mutually responsive to each other with a view on the (ethical) acceptability, sustainability and societal desirability of the innovation process and its marketable products (in order to allow a proper embedding of scientific and technological advances in our society'. Van den Hoven et al. (2013: 20) further connect RRI processes to the success of the products of innovation: 'consideration of ethical and societal aspects in the research and innovation process can lead to an increased quality of research, more successful products and therefore an increased competitiveness'.

To embed this proposed RRI framework in higher education, a focus is needed (1) on the scientists, in particular science students, as to what is required of them to become not only responsible researchers but 'responsible innovators' (Kallergi & Zwijnenberg, 2019), and (2) on the nature and scope of training afforded.

Almeida and Quintanilha (2017: 46) note that researchers require 'both the awareness of societal challenges and the ability of researchers to think about science in the broader context of society'. Tassone et al. (2017: 343), in considering RRI within the framework of the university and the grand challenges that science could address, extend this to 'fostering RRI in higher education curricula is about equipping learners to care for the future by means of responsive stewardship of research and innovation practices that address the grand challenges of our time in a collaborative, ethical and sustainable way'.

Several examples have been detailed with respect to the training of students to unlock such higher-order thinking (Heras & Ruiz-Mallén, 2017) required to contribute as RRI practitioners. The Higher Education Institutions and Responsible Research and Innovation project (HEIRRI) is a valuable resource for guiding such studies (HEIRRI, 2016).

In one embodiment, RRI anticipates the development of marketable products from research, requiring the training of students with a view towards the adoption of entrepreneurial mindsets. RRI also calls for science students to engage with the public in all aspects of the research and innovation pipeline, requiring, in turn, further training in engagement with the public.

The varied nature of the expectations of science researchers in the RRI framework represents a clear challenge to the university training of science students, necessitating cross-disciplinary approaches.

Biotechnology

Biotechnology is an applied field of study, drawing principally from the disciplines of engineering, chemistry and biology. Its simplest definition is the application of living organisms to produce new products, or to improve existing processes. Active research in biotechnology can be grouped into five areas of research applications: food, energy, water, the environment and health. New research in the five areas, including stem cells, drug discovery, wearable diagnostics, personalised healthcare, water treatment, biological energy, waste-water treatment, environmental remediation and even climate change, speaks to an area of scientific endeavour which directly influences many areas of human endeavour.

For modern science, the public turmoil around genetically modified organisms (GMOs) and the slow public acceptance thereof – fuelled by distrust, misinformation, sensationalism, corporate interest, as well as conspiracy theories over the past two decades – was unprecedented. It laid bare the disconnect between the public and role players such as industry, government and scientists in newer fields of scientific discovery. Correspondingly, it heralded a new era of public engagement with science, calling science and industry to account, squarely placing the field of biotechnology at the centre of revised approaches to science engagement internationally.

Public engagement with biotechnology in the South African setting sought to address scientific misinformation on several issues, through several science communication initiatives organised by the South African Agency for Science & Technology Advancement's Public Understanding of Biotechnology Programme. The focus was on deficit models of science communication, necessitated in part by broad divides in the public's access to education and information. While seeking to provide balanced information on science, the approach actively sought to showcase and highlight the benefits of biotechnology, while explicitly encouraging the adoption thereof as a future career for scholars.

Many countries view biotechnology as one of the cornerstones of scientific investment because of the aforementioned potential to impact so many areas of the lives of its citizens, as well as the economic leverage it may bring (OECD, 2009). Indeed, biotechnology is viewed as a hope for addressing some of the most pressing global challenges of our time (DST, 2013). The discipline's emphasis on applied research and product development means that the field also holds potential for entrepreneurship and for growing local economies.

In South Africa during the early 2000s, for example, several government-funded entities were created to oversee the funding and commercialisation of biotechnological research and products. A strong emphasis on the transfer of these technologies from research to commercial spaces called on universities to provide access to support and training for the development of entrepreneurship and technology transfer skills for its scientists. Similar to other countries, the aim is to encourage and provide support for 'academic entrepreneurs' (Miller et al., 2014) to commercialise research. In order to meet this demand, several entities such as the Technology Innovation Agency, the National Intellectual Property Management Office and the country's Department for Science and Innovation have sought to provide opportunities for non-curricular training in technology transfer and innovation.

Responsive to the role that the public holds in enabling scientific research to take place, research grant funding calls from the South African government (most notably the National Research

Foundation of South Africa) requires that grants clearly define the societal challenge that it would address, the application's alignment with national policies or strategies, and how the research outcomes could lead to addressing real societal challenges.

Increasingly, funding instruments in South Africa also call for more communication of scientific research to the public, while recent national policies (DST, 2007, 2013, 2015) in valuing the role of science communication, call for approaches that create a scientifically-literate society, viewing the public as a source of valuable insight into addressing localised problems.

As the above indicates, the call for greater involvement in science and research has multiple antecedents and enabling structures. Within the scope of biotechnology research in South Africa, the motivations for the study of RRI presented here include (1) the public being given a voice in decision-making around research and innovation processes; (2) science students (scientists) gaining a better understanding of the challenges faced by society in a specific area of research, while meeting and engaging the public for whom research is conducted; and (3) enhancing the public's role in science and technology, either the early acceptance or adoption of new technology by the public, or through the public providing localised perspectives on research, this form of engagement having the capacity to lead to the improved success of research products that are aimed at addressing societal challenges and improving the lives of the country's citizens.

Biotechnology research in South Africa's universities – guided by national policies to address societal issues such that it results in commercially viable products, in an academic climate that promotes active public–researcher engagement – resonates with core tenets of RRI. Viewed by others as 'a relevant and challenging case study for RRI' (Kallergi & Zwijnenberg, 2019), the field provides a specific context to explore the embedding of RRI into the training of postgraduate science students in biotechnology.

Co-creation, participatory research and citizen science

RRI has emerged as a focal point for public engagement in research, but few examples exist where public engagement in research has been applied in real scientific research. Given this vacuum, different models of public engagement such as citizen science, co-creation and participatory research are briefly explored here with respect to RRI.

Co-creation is a 'collaboration in which various actors actively join forces to tackle a shared challenge', in which priority setting and/or target setting are defined as part of the co-creation process (Vandael et al., 2018: 3). Co-creation principles are modelled on the equality of stakeholders in terms of their contributions, with stakeholders carefully considered in terms of their conception of a specific challenge and the tools that they bring to support successful co-creation (Vandael et al., 2018). The process can be limited by the time-consuming nature of this understanding of co-creation.

The UK's National Institute for Health Research (NIHR) funds partnerships between entities such as the NIHR Biomedical Research Centres and higher education, with a view to rapidly translating research from universities into innovative products that support patient needs (Greenhalgh et al., 2017). This initiative provides a real example of a 'value co-creation' model which seeks to involve patients in the design, delivery and dissemination of research needs (Greenhalgh et al., 2017).

A review of this model is underway to address challenges of relevance to RRI: the very nature of biomedical research innovation and product development that may neglect the priority setting of patients, as well as the reluctance by some scientists to fully engage with the public in all of these processes (Greenhalgh et al., 2017).

Conceptions of citizen science largely centre on citizens in a data-gathering role for a wide array of projects (Cohn, 2008). These include a wide range of topics, from monitoring bird sightings to amateur astronomers searching for interstellar dust (Hand, 2010). At least 60 000 volunteers are believed to be involved in a bird count that is at least 100 years old (Cohn, 2008). The information gathered is valuable and useful to science and many stories abound with respect to the value of discoveries made by citizen science. Undeniably, citizen science provides a route for science engagement with the public, for science learning (NAS, 2018) as well as encouraging involvement in science.

RRI calls for something fundamentally different to this conception of citizen science, premised on the meaningful input by non-scientists into the direction of research and the resulting innovation of products that can benefit their lives. RRI is not citizen science per se, but two factors see an intersection between RRI and citizen science.

RRI may be challenged by a lack of interest, insufficient knowledge or lack of trust in the process on the part of the general public to engage with scientists. Citizen science may indirectly provide a route to establishing relationships where communities have had prior engagement with scientists. As some researchers note, communities engaged in citizen science can lead to 'enhanced community science literacy' which may 'guide science in ways that advance community priorities' (NAS, 2018: 4).

Newer conceptions of citizen science extend the data-gathering role of citizen science beyond contributory and collaborative to co-creation, defining co-created projects as follows: 'the participants collaborate in all stages of the project, including the definition of the questions, development of hypotheses, discussion of results and response to further questions that might arise' (Senabre et al., 2018: 30, drawing from Follet & Strezov, 2015). Senabre et al. (2018) sought to address the lack of 'mechanisms' and tools available for enacting this mode of citizen science. Using existing mechanisms and facilitation tools for citizen science, the authors detail how 95 senior-school students and 5 scientists collaborated to design a 'citizen science research project' in a specific co-creation model (Senabre et al., 2018: 29). The core of this is the extension of citizen science into a model that draws from the principles of co-creation.

Community-based participatory research (CBPR) has been described as a 'collaborative approach to research that equitably involves all partners in the research process and recognises the unique strengths that each brings. CBPR begins with a research topic of importance to the community, has the aim of combining knowledge with action and achieving social change to improve health outcomes and eliminate health disparities' (Jull & Giles, 2017: 3, drawing from The Kellogg Foundation, 1992). CBPR resonates with an imagining of RRI processes that are pro-poor and committed to collaborating with 'marginalised communities' to address challenges identified by the community (Jull & Giles, 2017). In this embodiment, members of the community hold expertise and knowledge to help shape the research. CBPR as a process shifts the needle to equality between stakeholder communities and researchers, with the aim of ultimately leading to 'social transformation' of community members (Jull & Giles, 2017). A wide range of well-established CBPR tools such as participatory mapping, semi-structured interviews and focus groups are documented in the literature to support engagement between scientists and community members (Jull & Giles, 2017). CBPR holds elements of co-creation but allows for greater flexibility in the process, including in the numbers of community members engaged. A core benefit of CBPR is strengthening relationships at the scientist–society interface. In this respect, CBPR has been viewed as a valuable approach in sectors such as public health (Israel et al., 1998).

Biotechnology engagement models explored at Rhodes University

Against the backdrop of the scope of biotechnology nationally and internationally, Rhodes University's Biotechnology Innovation Centre (RUBIC) was formed in 2014 with the express purpose of providing an experimental, trans-disciplinary training space for postgraduate biotechnology students. The aim was to integrate biotechnology research and teaching, with courses in entrepreneurship as well as in science engagement. Four approaches for the incorporation of science engagement into the postgraduate training of biotechnology students were explored. In devising these approaches, the following was taken into consideration:

- The field of biotechnology is broad and while defined as an applied scientific discipline, certain students' research programmes were more fundamental in nature, precluding them from direct engagement with the general public. Projects and research programmes that were more readily applicable to peoples' lived experience were deemed preferable as we sought to develop the models. Consideration was given to research in areas of local and national prominence. Projects related to water treatment, alternative energy, sanitation and traditional medicines were identified.
- Research in biotechnology is frequently patentable. Any engagement with the general public should not compromise this intellectual property. Projects were also selected such that it did not hold the potential to infringe on any intellectual property of the stakeholders engaged.
- Engaging the public about enduring issues, such as medicines and health issues, could raise false hope of an immediate cure amongst impacted communities. Careful consideration of the ethics of engaging the public regarding certain research areas needed to be made.
- Many postgraduate students entering the biotechnology programme had no prior science engagement experience and were therefore not comfortable with engaging the public directly about their research without some form of training.
- Research which was very specialised, having a clear 'public' in mind, was viewed as an advantage. For example, existing interest groups allowed students to engage with a specific audience.
- A clear rationale for engaging the public in terms of the proposed benefit of the ultimate research needed to exist.
- For students, a programme had to be developed engaging

the public in a meaningful way such that their involvement enhanced the actual research or prototype development of ongoing research. In other words, engaging the public about their research needed to hold potential value to the students' research, to avoid it becoming a box-ticking exercise (a concern noted in other texts on the subject [Van Hove & Wickson, 2017]).

- Research engaging the public should have a legitimate question in mind, and seek to avoid interviewee fatigue.

Bearing the above in mind, the following models were examined as part of research into direct engagement between biotechnology science students and the general public:

1. Direct engagement between scientists and the public at a science fair;
2. Engage the public actively in laboratory-based research;
3. Engage the public about their views on new products; and
4. Engage specific publics regarding their perspectives on current and future research.

The focus in the first two models was on the specific benefit of the engagement to postgraduate science students, and the last two on the practical considerations of their application. The first model is discussed in some detail with respect to the benefits to science students, as part of a process in training students in RRI processes.

All research activities detailed received ethics clearance from Rhodes University's Ethical Standards Committee.

Direct engagement between scientists and the public at a science festival

This simple model takes advantage of existing opportunities for scientists to meet with the general public. Grahamstown – where this study was based – hosts Scifest Africa, a national annual science festival. The event which provided a vehicle for direct engagement was 'Speed-Date-a-Scientist' in which members of the public meet scientists either one-on-one or as part of a group for a short period of time, before the scientist moves to another group or individual.

Following this format of engagement, 15 biosciences (biotechnology, microbiology and biochemistry) postgraduate students were involved in a study detailed in a recent publication (Limson, 2018). This research wished to explore whether simple forms of engagement about scientific research (in general terms) would provide learning opportunities that would resonate with RRI learning outcomes.

Written and individual oral feedback from students showed a rich set of experiences in terms of benefits to students as scientists, with certain responses clearly linked to the higher-order thinking expected in RRI learning. Six key areas of benefit to students emerged, with students indicating that even this exercise in which they engaged with members of the public for a short period of time, and in which they identified as scientists, impacted on their communication skills, served as an affirmation of choice of career as a scientist, enhanced their motivation to conduct biotechnology research and helped shape their identity as scientists, and increased their confidence to act as scientists. Finally, some responses suggested that the engagement caused students to reflect on the nature of the research they do with a view to conducting research that benefited society. A detailed analysis of the feedback is provided elsewhere (Limson, 2018) and is summarised below.

Enhancing communication skills: Postgraduate students appeared to benefit from the engagement simply by improving on their communication skills. Their reflections on the experience also alluded to the fact that they reflected on how this could be extended to communicate clearly with other scientists.

Affirmation of choice of career as a scientist: During the engagement, students noted that viewing themselves through the lens of the high school learners (who largely comprised the members of the public participating in the event) resulted in a strong sense of affirmation regarding their choice of career.

Motivation within the field: In turn, students indicated a greater sense of motivation to continue in their field of research, in particular, the more senior students (PhD candidates).

Identity: Of interest to this study is the opportunity for introspection afforded to students in terms of their sense of identity after being placed in a position where they were viewed as scientists. Selected excerpts from Limson (2018) reflect this: [The engagement] 'forced me to question myself: "Am I a scientist"?'; 'When you are around scientists, it is normal and you don't think that you are any different, but when you are with the public, that is when you realised [sic] that you are a scientist'; 'When you speak to non-scientists you feel like a scientist'; 'It is only when you talk to the general public [that] you realise that you have acquired skills as a scientist'; 'Do I know what a scientist is and what a scientist does? I believe that a scientist [is] someone who introduces innovative solutions to current problems'.

Viewing themselves as scientists, they noted, enhanced their sense of value of themselves as scientists and their confidence to be and practise science. Excerpts from Limson (2018): 'It is good to see yourself as a scientist because it helps with your confidence as a scientist'; 'Made me reflect on what I knew and what I have achieved as a scientist'; 'It makes you feel needed and important'; 'It made me feel important'.

A surprising finding of this research was that deeper learning took place despite the brief nature of the engagement. Certain responses indicated that in coming to terms with their identity through self-reflection, some students also began looking outward and considered societal benefit and the real-world applications of their research. Students' responses: '[I am] more committed to [making] a difference in the community'; 'Speaking to budding scientists about subjects that interest me also affirmed my feeling that the science that I have chosen to be involved in is poised to make a difference in the world'; and 'I grew in confidence to do research that can be applied in the real world' (Limson, 2018).

Students also indicated that the engagement offered an opportunity to hear other points of view, a clear step towards RRI learning outcomes of true engagement between scientists and the public.

In order to further contextualise these responses, the study (Limson, 2018) used a framework generated by Heras and Ruiz-Mallén (2017) for the assessment of RRI learning outcomes.

Table 1 shows three of the four learning dimensions proposed by Heras and Ruiz-Mallén (2017), with a selection of the original associated outcomes, assessment criteria and indicators, detailed by the authors in their paper. (No indicators associated with the first learning dimension – basic cognitive aspects of learning – were included since indicators related to this were not present owing to the nature of the activity).

Feedback provided by learners (Limson, 2018) to the Speed-Date-A-Scientist were matched to different indicators as shown. A selection of these responses is reproduced in Table 1.

Linking feedback to indicators, outcomes and learning dimensions within the RRI framework provides a tool for researchers seeking to evaluate the nature of the anticipated RRI learning experienced by students.

The three learning dimensions shown in Table 1, in order of increasing complexity, with some associated learning outcomes are: experiential aspects of learning (the feelings and emotions, attitudes and perceptions experienced by students); transversal competencies (learning to learn, social and civic competencies, the sense of initiative gained); and RRI values (detailing emotional and cognitive engagement, critical and creative thinking) as proposed by Heras and Ruiz-Mallén (2017). Using this framework, the key outcome is the evidence of RRI learning as suggested by student feedback linked to indicators of RRI values detailed in Heras and Ruiz-Mallén (2017).

Table 1: Evaluation of RRI learning outcomes by using indicators and assessment criteria developed by Heras and Ruiz-Mallén (2017)

Learning outcome and/or process requirement*	Assessment criteria*	Indicator*	Evidence based on student feedback (selected examples)
	*Learning dimension: Experiential aspects of learning**		
Feelings and emotions	Enjoyment	Student's interest in science and learning science Excitement caused by science and learning science	'It was refreshing to speak about what I do in an informal manner'. 'Nice to get an opportunity to speak about your research. Generally your work does not get shared outside of a narrow community'. 'Seeing their passion reminds you of yours...' '[It] motivates me to carry on. The interest and amazement feeds your own passion and motivation to carry on in your field'.
Feelings and emotions	Emotional awareness and reflexivity	Student's ability to reflect upon and through her/his emotional responses and make consistent behavioural choices in the activity	'[I] feel like I am representing the scientific fraternity'. 'As I speak to people, I want to be credible and that motivates me to do my best in the lab'.
Feelings and emotions	Empowerment and sense of belonging	Student's sense of belonging to the community when doing the scientific activity Student's feeling recognised by other participants beyond their classmates	'It was like looking at yourself in the mirror, talking to yourself ten years ago'. 'Having someone else appreciate your work makes you see your work through their eyes'. 'Am I a scientist? Why do I do what I do?' 'It never crossed my mind that I am a scientist. It's only when you meet people who are not exposed to science that you realise that you are a scientist'.
Attitudes and perceptions	Perceptions of science and the scientific issues approached	Student's perceptions of scientists, scientific careers and/or jobs	'It is only when you talk to the general public [that] you realise that you have acquired skills as a scientist. It is quite enlightening'. 'When you are around scientists, it is normal and you don't think that you are any different, but when you are with the public, that is when you realised that you are a scientist'.
Attitudes and perceptions	Attitudes towards science and the scientific issues approached	Student's curiosity and interest towards science Student's interest in scientific careers and/or jobs	'The [high school learners'] enthusiasm for what I do made me feel more [certain] about my choice to do biotechnology'. The activity 'inspires you to continue [in your field]'. 'Engaging with [the high school learners] and teaching them about my research allowed me as a scientist to share the knowledge and also re-ignited my passion for science'. 'Helps me appreciate more what [scientists] do'.

Learning outcome and/or process requirement*	Assessment criteria*	Indicator*	Evidence based on student feedback (selected examples)
		**Learning Dimension: Transversal competencies*	
Learning to learn	Understanding the value of learning	Student's awareness of the professional value of learning science Student's satisfaction to be able to learn science	'Talking and explaining to the [high school learners] I felt was quite inspiring as it reminded me of my purpose as a scientist and why I got into this research'. 'Made me feel grateful for the opportunity to be a scientist'.
Learning to learn	Reflective thinking	Student's reflection on her/his own learning during the activity	'Engaging with the high school [learners] helped me to understand my project even better'. 'Being able to communicate with the public helps you to communicate better to other scientists'. 'It is only when you talk to the general public [that] you realise that you have acquired skills as a scientist. It is quite enlightening'.
Social and civic competencies	Communication skills	Student's ability to elaborate and share ideas verbally and written during the activity	'The event provided an opportunity for self-reflection with regards to my ability to communicate with the public as a "scientist"'. 'I feel that by taking part in the speed dating [event], I also learned a bit more about how I could talk about science as I myself was more relaxed in the environment and found it easier to try and simplify things'. 'Being able to communicate with the public helps you to communicate better to other scientists'.
Sense of initiative	Entrepreneurship	Student's belief in her/his own ability to perform a scientific activity	'Made me reflect on what I knew and what I have achieved as a scientist'. '[I] feel like I am representing the scientific fraternity'. 'As I speak to people, I want to be credible and that motivates me to do my best in the lab'.
Sense of initiative	Self-confidence and esteem	Student's belief in her/his own ability to do well in a scientific domain Student's belief in her/his own verbal ability to discuss about science	'I feel that by taking part in the speed dating [event], I also learned a bit more about how I could talk about science as I myself was more relaxed in the environment and found it easier to try and simplify things'. The activities 'made me grow as a person, made me feel comfortable to rely on my own ideas [and to explore those as a scientist]'.
		*Learning dimension: RRI values**	
Engagement	Emotional engagement	Student's feelings when experiencing the activity, if any Student's further interaction and initiatives related to the activity once it is over	'[It] motivates me to carry on. The interest and amazement feeds your own passion and motivation to carry on in your field'. 'Interacting with eager [high school learners] who were curious about careers in science was especially motivating'.

Learning outcome and/or process requirement*	Assessment criteria*	Indicator*	Evidence based on student feedback (selected examples)
Engagement	Cognitive engagement	Student's ability to develop ideas and engage in higher-order thinking Student's willingness to continue working on the activity out of class	'Do I know what a scientist is and what a scientist does? I believe that a scientist [is] someone who introduces innovative solutions to current problems'. 'I would not mind participating in other events that are similar to this one because such events are very helpful in improving scientific communication skills to different audiences'.
Critical and creative thinking	Connecting topics with experience	Contextualisation of scientific topics within societal challenges in the activity Use of student's previous experiences and knowledge as a basis for learning in the activity	'I am personally motivated by research that could be beneficial to people'. [I feel] 'more committed to [making] a difference in the community'. 'It further reinforced the relevance of the work that scientists do and I saw that by observing the eager response of the [high school learners] while explaining different aspects of my work and the work that is done in my lab'.
Critical and creative thinking	Seeking other points of view	Student's ability to consider different perspectives and points of view	The 'science engagement activity also provices the opportunity to scientists not only to educate but to listen and learn from the public'.

* Selected Learning Dimensions, Outcomes and Assessment Indicators listed here are extracted from Heras and Ruiz-Mallén (2017)

Table reproduced from Limson (2018) in part and drawing from Heras and Ruiz-Mallén (2017). Evidence is based on selected student feedback drawn from Limson (2018).

Engage the public actively in laboratory-based research

The second model actioned in the centre sought to actively engage the public in a meaningful way such that their involvement either enhanced ongoing research or prototype development. The example described below was selected since it sought to address real issues related to water treatment and alternative energy generation, both contemporary and enduring concerns that most publics in South Africa can relate to.

A biotechnology master's student invited non-scientists to assist her in conducting experiments linked to her research project over a two-day period. The masters student's research was centred on microbial fuel cell devices for waste-water treatment. These fulfil dual roles: they both treat a range of different waste waters, and by utilising bacteria, are able to generate small amounts of direct electricity. The student developed miniature models of the microbial fuel cells and wished to establish the ruggedness of the basic design when operated by non-scientists. Establishing this was of relevance in terms of future scale-up of the miniaturised microbial fuel cells to allow for treatment of larger volumes of waste water.

A detailed analysis of this study will appear elsewhere. Briefly, feedback from the biotechnology student after the engagement yielded similar themes to the first approach detailed above, including affirmation, motivation, identity and the beneficial impact that engagement had on her own ability to communicate with the public. Feedback indicated that the engagement helped her reflect on why she entered science and helped her understand her own work better. She noted how the response ('excitement') of the non-scientists to being involved in real scientific experiments motivated her in her research, calling the experience 'energising'. Feedback provided indicated higher-order thinking linked to RRI values (Heras & Ruiz-Mallén, 2017) not observed in the Speed-Date-a-Scientist activity, and is linked to the greater length of time and greater depth of the engagement. The student twice referenced 'responsibility' as in a 'renewed responsibility' as a scientist as well as the 'burden of responsibility' on scientists for honesty. Of specific interest to the RRI values espoused by several authors is her reflection that the engagement reminded her of the need for scientists to be ethical in their research. This was an important outcome of this engagement and can be associated with both the greater length of time and the greater depth of the engagement. The public were participants who could support the research outcomes, and were viewed as valuable to the research itself.

Also in line with this being research aiming for future product development, the students reflected that it provided an impetus for her to commercialise the outcomes of her research.

The student notes that bringing members of the public into the research laboratory provided opportunities for the public to collaborate with scientists and that it could provide opportunities for the public to help shape the direction of research.

Engage the public about their views on new products

A third model sought to conduct user surveys to gain localised perspectives from communities at the end of the fundamental research, but at the start of prototype development to enhance its potential for adoption. One such example is summarised here.

Research in the field of nanotechnology has adopted the approach that, while it is understood that there may be health concerns using materials as small as a nanometre, that research into its potential environmental fate and impact on human health would continue alongside research into studying the properties of these materials for addressing major scientific challenges.

One such challenge is the purification of water. Nanofibres, materials with a diameter in the nanometre range, can be produced from a range of different materials, mostly from different polymers in a process known as electrospinning. These materials, with their high surface area to volume ratio, offer a wide surface area for the adsorption of contaminants in water, acting as an effective 'sponge'. When coated with different materials, the power of the nanofibres to remove and inactivate bacteria can be enhanced.

Research in our laboratory has developed a nanofibre-based process that removes both metals and bacteria, common contaminants in drinking water from municipal supplies where conventional treatment processes have failed. Before turning this into a product, research sought to engage communities about their specific needs for a device at home that could treat water at the point of use, as well as their thoughts on the design thereof. The results of this study will be detailed elsewhere.

In brief, community members were willing to talk to the researcher given the local problem of water in the Makana Municipality area. Several indicated a willingness to field-test a final prototype. Suggestions for the ultimate design varied, based on access to water infrastructure, indicating that two different designs were required to meet different community challenges.

Engage specific publics regarding their perspectives on current and future research

The RRI framework calls for the engagement of the public at the earliest stages of scientific research. An attempt to explore this approach within the traditional medicines sector is briefly described here.

In a collaboration with the Rhodes University Faculty of Pharmacy, our research sought to explore the safety, variability and potential toxicity of traditional medicines that had been prescribed for common ailments such as water-borne diarrhoeal disease. This research was conducted against the backdrop of new laws for regulation of traditional medicines. Our research wished to establish, in part, perceptions regarding the testing of traditional medicines. Both potential consumers of traditional medicines as well as traditional health practitioners were interviewed in this study.

Briefly, of relevance to the broader scope of RRI, this research highlighted the need for an understanding of cultural and religious beliefs which impact the public's perceptions. For example, while a scientific basis may exist to test traditional medicines (for example, the fact that seasonal variation or different soil conditions may alter the concentration of pharmacologically active ingredients in plants used in traditional medicines), traditional health practitioners indicated that the spiritual dimension of traditional healing cannot be tested by scientific methods. Feedback from some users of traditional medicine indicated that their belief in the efficacy of traditional medicine is based on trust and culture. Therefore, while the establishment of scientific testing could in fact be offered in order to test the efficacy of traditional medicines, the adoption and use of such services, in particular when prescribed by a traditional health practitioner, may be limited to some extent.

Lessons learnt and conclusions

The lack of accessible models for implementing RRI into university research represented both a challenge in terms of a lack of benchmarking, but also an opportunity to develop engagement processes that, on the one hand, simply supported biotechnology research as well as student learning. On the other, the approaches described provided separate and different opportunities to engage the public in research.

Students in RUBIC are all provided with opportunities to engage in science communication, through writing, video production or animation courses. Idea generation, entrepreneurship training and business planning courses provided, aim to do more than tick the boxes to encourage entrepreneurship. In these spaces, it was hoped, students would, in imagining, meeting or directly engaging non-scientists, evolve an understanding of the communities their research would ultimately impact. Equipped with a unique set of skills for students, feedback from students involved in this study indicated the untapped potential that both simple and advanced forms of engagement with the public hold for the development of their own personal identity, confidence, motivation as scientists and a desire to conduct research for societal benefit. The unlocking of higher-order RRI thinking of ethics, honesty, responsibility to the public and more, yielded to a desire to turn such research into real, marketable solutions that engages and benefits the public.

While research at RUBIC in the study of science engagement with the public at RUBIC is in its infancy, the study identified some challenges and opportunities that scientists may encounter in public engagement in RRI.

Language is a challenge especially in countries such as South Africa with eleven official languages. While there may be no available translation of scientific terms into different languages, students' ability to improvise and to speak in the mother tongue of those being interviewed, was an enabling feature of the engagement. As one student noted, stakeholders could look beyond the scientist in front of them and connect with the human being and the message of the research when communicating in the mother tongue of the public being interviewed.

Specific research topics may be limited in scope, and matching specific publics to areas of research interest was key (e.g. issues on water and sanitation). For communities to engage researchers, trust and relationship-building play an important role. For some of the activities described, existing relationships between academics at Rhodes University supported research. Certain community groups resisted engaging with researchers from universities, citing past experiences where researchers were insufficiently prepared to engage in a manner which was culturally sensitive. In this scenario, trusted intermediaries both schooled student researchers involved in the research identified above and mediated the engagement.

Co-creation in its strictest sense would have limited application in the form of engagement discussed here, but holds value for small working groups representing different parts of the so-called 'triple/ quadruple helix', engaging carefully identified members of the public, scientists, government and industry in problem identification. The broader mandate of science and, in particular, government policy in South Africa, envisions problem-setting in a wider space, engaging a broad transect of communities and community members, capitalising on such engagements to provide a clearer understanding of research challenges that science could address. The engagement also envisions the benefits of shaping a more scientifically literate society amongst participating members.

Central to all engagement activities with the public were issues of trust and relationship-building. A clear limitation for these studies was the lack of training in tools for societal engagements. Studies in our research group are currently exploring partnerships with social scientists skilled in the tools of CBPR as a basis for future engaged research activities that may support relationship-building.

As stated before, RRI aims for research to be conducted with and for society. It also calls for the commercialisation of research from protectable intellectual property. While students are well-versed in aspects of what constitutes disclosure, clear discussions with the public should ideally delineate areas of questioning such that it protects indigenous knowledge.

There exists the potential for conflict of interest – engaging the public in the research and innovation process may compromise trust between researchers and the public if such research is commercialised (Miller et al., 2014). Clearly defining the confines of interviews or engagements can help build trust between researchers and the public. Miller et al. (2014) argue for the need to address potential commercialisation early on during public engagement. Biopiracy and theft of indigenous knowledge has contributed in part to a culture of distrust between holders of traditional knowledge and researchers, reinforcing the need for clear discussions about rights to intellectual property, as one of the steps to establishing longer-term relationships.

Another consideration to bear in mind is the source of research funding in biotechnology. Caulfield et al. (2006) highlight how the credibility of researchers in the field of biotechnology declines if they are funded through industry rather than government. Other studies suggest that this deficit of trust is based on the perception of motivations, government-funded research being associated with benevolence rather than self-interest (Caulfield, 2006, drawing from Critchley, 2008).

Recent studies suggest that knowledge of RRI as a policy is low amongst scientists. However, notions of responsibility to do sound, 'publicly legitimate research' exist (Glerup et al., 2017). The challenges of RRI call for 'a more rigorous contribution of the humanities to science and technology education' (Kallergi & Zwijnenberg, 2019).

Future research will focus on partnerships with social scientists to assist in training of researchers in public engagement, with a view to offering deeper analyses and outcomes. Research will explore the perceived benefits of engagement from the viewpoint of the general public, to help establish a clearer picture of the value of the approaches adopted here for public engagement in research.

Acknowledgements

Funding for this research was provided through the DST/NRF South African Research Chair in Biotechnology Innovation & Engagement (grant number 95319). The following collaborators are acknowledged for their collaboration in the science engagement approaches detailed here: Prof. Roman Tandlich and Priscilla Keche for research in traditional medicines, Aphiwe Mfuku and Dr Ronen Fogel for research in water purification testing, Dr Ronen Fogel for collaborating on Speed-Date-a-Scientist activities, and the staff and students of RUBIC for participating in the range of science engagement activities described.

References

Almeida, M. S. & Quintanilha, A. (2017). Of responsible research: Exploring the science–society dialogue in undergraduate training within the life sciences. *Biochemistry and Molecular Biology Education*, 45(1), 46–52.

Caulfield, T., Einsiedel, E., Merz, J. & Nicol, D. (2006). Trust, patents and public perceptions: The governance of controversial biotechnology research. *Nature Biotechnology*, 24, 1352–1354.

Critchley, C. R. (2008). Public opinion and trust in scientists: the role of the research context, and the perceived motivation of stem cell researchers. *Public Understanding of Science*, 17(3), 309–327.

Cohn, J. P. (2008). Citizen science: Can volunteers do real research? *BioScience*, 58(3), 192–197.

Department of Science and Technology (DST) (2007). *Innovation towards a Knowledge-Based Economy: Ten-year plan for South Africa (2008–2018).* http://www.sagreenfund.org.za/wordpress/wp-content/uploads/2015/04/10-Year-Innovation-Plan.pdf.

Department of Science and Technology (DST) (2013). *The Bio-Economy Strategy.* http://www.naci.org.za/nstiip/index.php/knowledge-base/stratergies/13-bio-economy-strategy.

Department of Science and Technology (DST) (2015). *Science Engagement Strategy.* http://www0.sun.ac.za/scicom/wp-content/uploads/2018/06/2015_sci_engagement_strategy.pdf.

European Commission (2019). Responsible research and innovation. European Commission website. https://ec.europa.eu/programmes/horizon2020/en/h2020-section/responsible-research-innovation.

Follett, R. & Strezov, V. (2015). An analysis of citizen science-based research: Usage and publication patterns. *PloS One*, 10(11), e0143687. doi:10.1371/journal.pone.0143687.

Glerup, C., Davies, S. R. & Horst, M. (2017). 'Nothing really responsible goes on here': Scientists' experience and practice of responsibility. *Journal of Responsible Innovation*, 4(3), 319–336.

Greenhalgh, T., Ovseiko, P., Fahy, N., Shaw, S., Kerr, P., Rushforth, A., et al. (2017). Maximising value from a United Kingdom Biomedical Research Centre: Study protocol. *Health Research Policy and Systems*, 15. doi: 10.1186/s12961-017-0237-1.

Hand, E. (2010). Citizen science: People power. *Nature*, 466, 685–687. doi: 10.1038/466685a

Heras, M. & Ruiz-Mallén, I. (2017). Responsible research and innovation indicators for science education assessment: How to measure the impact? *International Journal of Science Education*, 39(18), 2482–2507.

HEIRRI (2016). Deliverable 2.2 State of the art review. http://www.guninetwork.org/files/images/imce/heirri_wp2_d2.2.pdf.

Israel, B. A., Schulz, A. J., Parker, E. A. & Becker, A. B. (1998). Review of community-based research: Assessing partnership approaches to improve public health. *Annual Review of Public Health*, 19, 173–202.

Jull, J. & Giles, A. (2017). Community-based participatory research and integrated knowledge translation: Advancing the co-creation of knowledge. *Implementation Science*, 12. doi: 10.1186/s13012-017-0696-3.

Kallergi A. & Zwijnenberg, R. (2019). Educating responsible innovators-to-be: Hands-on participation with biotechnology. In A. Reyes-Munoz, P. Zheng, D. Crawford & V. Callaghan (eds), *EAI International Conference on Technology, Innovation, Entrepreneurship and Education* (pp. 79–94). TIE 2017. doi: 10.1007/978-3-030-02242-6_7.

Kellogg Foundation (1992). *Community-Based Public Health Initiative*. Battle Creek, MI: Kellogg Foundation.

Limson, J. (2018). Putting responsible research and innovation into practice: a case study for biotechnology research, exploring impacts and RRI learning outcomes of public engagement for science students. *Synthese*. doi: 10.1007/s11229-018-02063-y.

Miller, F., Painter⊠Main, M., Axler, R., Lehoux, P., Giacomini, M. & Slater, B. (2014). Citizen expectations of 'academic entrepreneurship' in health research: Public science, practical benefit. *Health Expectations*, 18. doi: 10.1111/hex.12205.

National Academies of Sciences, Engineering, and Medicine (NAS) (2018). *Learning Through Citizen Science: Enhancing opportunities by design*. Washington, DC: The National Academies Press. doi: 10.17226/25183.

OECD (2009). *The Bioeconomy to 2030: Designing a policy agenda*. Paris: OECD.

Owen, R., Macnaghten, P. & Stilgoe, J. (2012). Responsible research and innovation: From science in society to science for society, with society. *Science and Public Policy*, 39, 751–760. doi: 10.1093/scipol/scs093.

Ribeiro, B. E., Smith, R. D. J. & Millar, K. (2017). A mobilising concept? Unpacking academic representations of responsible research and innovation. *Science and Engineering Ethics*. doi: 10.1007/s11948-016-9761-6

Senabre, E., Ferran-Ferrer, N. & Perelló, J. (2018). Participatory design of citizen science experiments. *Comunicar*, 26. doi: 10.3916/C54-2018-03.

Tassone, V. C., O'Mahony, C., McKenna, E., Eppink, H. J. & Wals, A. E. J. (2017). (Re-)designing higher education curricula in times of systemic dysfunction: A responsible research and innovation perspective. *Higher Education*, doi: 10.1007/s10734-017-0211-4.

Van Hove, L. & Wickson, F. (2017). Responsible research is not good science: Divergences inhibiting the enactment of RRI in nanosafety. *NanoEthics,* 11(3), 213–228.

Van den Hoven J., Nielsen L., Roure, F., Rudze, L., Stilgoe, J., Blind, K., et al. (2013). Options for strengthening responsible research and innovation. European Commission Report. https://ec.europa.eu/research/swafs/pdf/pub_public_engagement/options-for-strengthening_en.pdf.

Vandael, K., Dewaele, A., Buysse, A. & Westerduin, S. (2018). *ACCOMPLISSH Guide to Co- Creation*. Ghent: Ghent University.

Von Schomberg R. (2012) Prospects for technology assessment in a framework of responsible research and innovation. In M. Dusseldorp & R. Beecroft (eds), *Technikfolgen Abschätzen Lehren* (pp. 39–61). Wiesbaden: VS Verlag für Sozialwissenschaften.

3 Putting responsible research and innovation into practice at a local level in South Africa

Penelope S. Haworth & Anne M. Dijkstra

Introduction

In a small volume entitled *Science and Survival*, published in 1966, Barry Commoner, then Professor of Botany at Washington University, begins his discourse with the question 'Is science getting out of hand?' Commoner explored many of the issues with which society is grappling in the second decade of the 21st century. Not least of these is a strident lack of trust between science and the society it purportedly serves. The concerns are not new: current issues resonate through chapter headings such as 'Science versus society', 'The ultimate blunder', 'The scientist and the citizen' and finally 'To survive on earth'. Tellingly, he uses terminology such as 'the erosion of science's integrity' (pp. 60–61), 'agricultural devastation' (p. 73), the 'assault on the biosphere' (p. 75). What is clear is that for at least the last 50 years, since the very obvious devastation and salutary lessons of the Second World War, people have been aware that the planet's 'thin life-supporting surface' (Commoner, 1966: 110) is under siege. Yet, exponential population growth, industrial and technological development and rampant consumerism have continued without any real consideration of their effect on a finite and finely balanced biosphere.

As addressed by Cochrane, Sauer and Aswani (2019) working in the field of coastal and marine science in South Africa, the world is facing social and environmental challenges such as ensuring sustainable use of resources and safeguarding biodiversity. They argue that to address modern-world challenges, changes in South African attitudes – and broader – are needed. Their study of presentations at the 2018 South African Marine Sciences Symposium (SAMSS) shows, however, that very few of the presentations from the coastal and marine sciences community could be assessed as actionable or directly relevant to societal needs (Cochrane et al., 2019: 4).

The recent White Paper on Science, Technology and Innovation (DST, 2019) published by the South African government also acknowledges this rapidly and fundamentally changing world. Drivers for these global changes are socio-economic and geopolitical, scientific and technological, and environmental. The White Paper sets the long-term policy direction for the South African government with the aim for a more prosperous and inclusive society via a growing role for science, technology and innovation. It suggests policy approaches which include developing ways to support the knowledge enterprise, and a role for science engagement and science communication. To make changes in South Africa possible, according to the White Paper (2019), society will need to value science, appreciate the impact of innovation on development, and anticipate and plan for change. Then, the potential of science, technology and innovation will be developed and advance South Africa.

Important in this policy, therefore, is that the needs of society will be taken into account. More specifically, to be able to develop a knowledge-based society and a healthy economy, South Africa should develop a responsible research and innovation (RRI) approach which includes, amongst others, a role for science engagement and communication (DST, 2019).

In this chapter, we explore how science engagement and communication can contribute to putting RRI into practice in South Africa and, consequently, assist in aiming for a more prosperous and inclusive society. We begin by providing a description of RRI and how it is embedded in South Africa. We then discuss experiences of implementing RRI through science engagement and communication in a South African research institute. We base our findings and experiences on the results from a European-funded H2020 project – NUCLEUS – to gain insights from the achievements and challenges for science engagement and communication in developing South African society. The chapter ends with a discussion and conclusions.

Responsible research and innovation (RRI) in perspective

In examining responsible research and innovation (RRI), Rip (2014: 1) refers to it as 'a social innovation' which 'catapulted from an obscure phrase to an issue in the European Commission's Horizon 2020 Program'. In recent years, the concept of RRI has been increasingly addressed in academic literature (e.g. Rip, 2014; Shelley-Egan et al., 2018). Burget et al. (2017) argue that the concept is still in development. According to Rip (2014) and Shelley-Egan et al. (2018), ideas about responsible innovation – then not yet labelled as RRI – developed, amongst others, from a report by the British Royal Society and the Royal Academy of Engineering (RSRAE, 2004) which discussed nanotechnologies and possible strategies for dealing with them in the future. In this report, the promotion of a wider dialogue about emerging technologies was also proposed as well as ways of implementing such a dialogue in practice.

Rip (2014: 2) explored the position of RRI in what he terms 'a historically evolving division of moral labour' as the roles and responsibilities of 'actors and stakeholders in research and innovation' are articulated and developed. Accordingly, scientists can no longer leave it to others to consider social, ethical and political issues. It is clear that in an increasingly global context, scientists and citizens need to work together.

Definitions of RRI emphasise the inclusion of all societal actors in the process of aligning research and innovation outcomes to the needs and expectations of society. For example, Von Schomberg

(2013: 19) defines RRI as a 'transparent and interactive process by which societal actors and innovators become mutually responsive to each other'. Meanwhile, the European Commission (EC) understands RRI as an inclusive approach to research and innovation which ensures that societal actors work together during the whole research and innovation process. In their view, RRI aims to better align both the process and the outcomes of research and innovation, with the values, needs and expectations of European society (European Commission, 2017). In practice that means, according to the European Commission, designing and implementing policy that will engage society in research and innovation developments; increase access to scientific results; ensure gender equality both in the research process and in the research content; include the ethical dimension and promote formal and informal science education. These aims have been translated by the EU into six key areas where RRI can be put into action: governance, public engagement, open access, gender equality, ethics and science education.[1]

In the South African approach to RRI, articulated in the White Paper (DST, 2019), the influence of these six key areas is clearly visible, viz.: (i) engagement of all societal actors throughout the process of framing societal challenges and developing joint solutions; (ii) addressing racial and gender transformation to unlock the full potential of South African society; (iii) improving the educational and skills profile of South Africans; (iv) increasing open access to science, technology and innovation (STI); (v) maintaining a high level of ethics in terms of the relevance and acceptability of STI to society and environmental sustainability; and (vi) developing the required governance framework to drive the RRI agenda across the National System of Innovation (NSI).

In the next section, we will provide findings about South Africa from the NUCLEUS project, which aimed to bring RRI to life in universities and research institutes in various countries. The

1 https://ec.europa.eu/programmes/horizon2020/en/h2020-section/responsible-research-innovation

findings from the NUCLEUS project will serve as a case study of RRI in action. More specifically, in describing how the key areas are brought into practice, the role science communication and engagement play in fostering a responsible science–society relationship will be described.

Finding fertile ground for embedding RRI

NUCLEUS, a four-year project funded by the European Union through the Horizon 2020 programme, ran from 2015 to 2019. The acronym stands for New Understanding of Communication, Learning, and Engagement in Universities and Scientific Institutions. Basing its definition of RRI on the definition by Von Schomberg (2013) as described above, the project aimed to gather a broader cultural, international and enriched perspectives on what a responsible science–society relationship entails. Therefore, in the first phase of the project, the way RRI is shaped in various situations was analysed. In the second phase, based on the roadmap extracted from the recommendations from the first phase, elements of RRI, for example, regarding public engagement and science communication, were implemented at ten universities and scientific institutions. In addition, activities to foster RRI were organised in various other places and spaces. Below, we will present lessons learned from both phases.

First phase: Identifying a broader perspective on RRI

In the first phase of the project, RRI was explored by means of conducting various studies. This included field trips, each of which took one particular perspective to find out how RRI was embedded in diverse contexts. The field trip to South Africa took the perspective of civil society (Doran, 2016). The trip was facilitated by the South African Agency for Science and Technology Advancement[2] (SAASTA) which is the country partner on the

2 http://www.saasta.ac.za/

NUCLEUS Consortium and the driver of the project in South Africa on behalf of the National Research Foundation. Visits were paid to SAASTA, the Osizweni Education and Development Centre and the National Zoological Gardens in Pretoria. At each location interviews were conducted with various members of civil society organisations such as science centres, community groups, education governance officials, teachers, business, zoos and others.

From these interviews the following observations were made (see Doran, 2016). The interviews revealed enthusiasm for engagement with civil society among museum staff and educators. Despite that enthusiasm, respondents indicated that engagement was mainly possible when tasks were within job roles and dedicated budgets were available. Interviews also showed that diversity and access to education is a challenge for various groups. Science centres in South Africa provide an outlet for informal learning and offer access to facilities for some schools further away from universities, but they are also in need of funding and equipment. Interviews with learners showed that they saw possibilities for their career paths via participation in activities offered by science centres. The question is how existing relationships between universities, science festivals, communities and organisations such as SAASTA can be taken to the next level to embrace RRI. A significant challenge that may prove to be a barrier to implementing RRI is funding.

On the other hand, there is also good opportunity to engage with civil society through citizen science projects, as is demonstrated with the Cradle of Humankind where communities and researchers connect with mutual learning benefits as outcomes. In this project, researchers, from South Africa and abroad, worked together with cavers and members of the local community on the discovery of a new species of a human relative, *Homo naledi,* in Maropeng. It included an open approach to social media and a coordinated communication effort that led to global coverage of the discovery and the research. The University of Witwatersrand played a role in convincing the collaborating parties that the story belonged to humanity as a whole and not to a single news network, and that the discovery should be shared globally. The interviews highlighted how researchers can work together with the local and global community in a research project. To ensure a long-lasting relationship, those involvements should always be mutually beneficial (Doran, 2016).

Next, to include an intercultural context of RRI, a cultural adaptation study was conducted (Dijkstra et al., 2017) while for the European perspective interviews were conducted with European researchers (see Böger, 2017, not reported on here). The cultural adaptation study included the cases of China and South Africa. Research questions for the cultural adaptation study focused on how RRI and other related concepts are implemented in international contexts; what barriers and successes affect the future implementation of RRI; and what can be recommended for the future implementation of RRI in universities and research institutes (Dijkstra et al., 2017).

For data collection for the cultural adaptation study, a multi-methodological and qualitative approach was applied. The use of various qualitative methods allowed for more insightful understanding and a broader cultural perspective on RRI (cf. Patton, 2002). However, there are also limitations since qualitative research can never be statistically representative and the results should be seen from that perspective. Both a literature study and interviews were conducted. The literature review included multiple sources of information, such as academic literature, reports, news articles, but also policy documents, statistical reports and personal communication. Semi-structured interviews were conducted with the aim of gathering further insights into practices in both countries. The protocol for the interviews was based on the questions from the European interviews and adapted after testing. Questions probed for background information; challenges for research and society; engagement; impacts of research on society; governance of research; changes foreseen in current practices and policies; responsibilities; and support wanted or needed. As a final question, respondents were asked what they expected from Europe regarding RRI.

In total, for the South African study, 13 interviews were

conducted, either via Skype or face-to-face, and recorded. The recordings served as the basis for analysis which took place at the conceptual, governmental or political, institutional and individual level. Respondents, who were asked for informed consent, held various leading positions in universities and science centres as leading researchers, university or faculty management, management or senior officers. Of those interviewed, 12 were male, 1 was female, and their ages ranged from 38 to 75 years.

RRI in South Africa at the governmental, institutional and individual level

At the governmental level, innovation is seen by both the South African government as well as interviewees as a means to advance the economy and lives of people. Programmes for technology innovation and research support are in place both for basic sciences as well as for strategic areas. Promotion of public engagement is included in these programmes. The science system, according to the interviewees, although one of the best in the region, faces challenges, such as funding which influences research output. In addition, access to universities has become more difficult for those with fewer financial means due to higher tuition fees. Equal inclusion to research and innovation regarding both gender and those from different population groups has the attention of government. However, according to some interviewees, a difference is reported for equal access in practice due to poverty and affordability of university education. Policies stimulate collaboration between indigenous knowledge holders, practitioners and researchers and industry. Various collaborations exist, for example, where the San people are working with industry on the *kougoed* plant (*Sceletium tortuosum*), which may be seen as a form of engagement and an application of RRI in practice. In the research process, San people have a say in what research is conducted and how, which shows bottom-up engagement.[3] Engagement efforts are also part of policy objectives of the Department of Science

3 https://mg.co.za/article/2015-02-19-bushmen-cure-all-offers-locals-a-sustainable-income

and Technology (DST). Through the agency of SAASTA the DST provides funds for science education and outreach which are allocated to foster awareness about science and technology. Effects, however, are difficult to measure (Dijkstra et al., 2017).

At the institutional level, SAASTA plays a major role in fostering efforts for science education and outreach, for example, by distributing materials, organising competitions or exhibitions and science festivals and providing training. Various science centres are funded by the DST. Rural areas are less developed than cities and are hard to reach for science education or outreach activities. Funding issues impact the ability of universities and other institutions to perform such tasks with limited means. South Africa is leading in open access policies (Unesco, n.d.) and these policies are taken up by several universities and institutes. The National Research Foundation, the main funding research agency in South Africa, considers impact and engagement to be important for the success of research projects. However, impact and engagement are not formalised in the key performance indicators applied to determine the success of research projects, so uptake by researchers is understandably limited as many feel that there is no tangible benefit for them. In addition, research proposals should adhere to ethical standards.

In practice, at the institutional level, the social impacts of research, as well as environmental impacts of research and innovation, appear to be considered as most important (Dijkstra et al., 2017). At the individual level, it was observed that researchers as well as science educators are performing many tasks with limited means. Engagement or outreach are not always considered part of their job but may be stimulated via role models. Inclusion, such as equal access to universities and research positions, are topics of concern for interviewees. Re-addressing existing differences will need careful strategies, they emphasised. Also, they considered equality to be an important aspect of the science–society relationship which may enhance trust and needs openness, transparency, respect and balance. Organising and participating in outreach and science education activities which may help development and engagement was also seen as valuable. In addition, being considerate and respectful towards citizens and participants in research was an attitude shared by many interviewees (Dijkstra et al., 2017).

To conclude from the cultural adaptation study, at the conceptual level, in South Africa, the terminology of RRI is not yet well-known. However, this does not mean that the ideas behind it or the elements of RRI are unknown to South African researchers. On the contrary, there are many instances where efforts can be seen as RRI in action and that show that RRI is put into practice. Some elements were more prominent than others. Equality, science education, and outreach are most developed and present at the governmental, institutional and individual levels. Open access is less prominent and is seen primarily at the institutional and individual level. Stakeholder and public engagement, as well as attention to the potential broader impacts of research and technology – and being responsive to stakeholders, the public or potential impacts – are less prominent. Ethics are seen as important, but the main focus of researchers is on doing their job and not on ethical reflection. According to the findings from the study, the South African interpretation of RRI focuses mainly on equality and science education and outreach. Other elements are present, but to a lesser degree and, in the case of assessing the broader impacts of research, not perceived to be equally relevant for fundamental research as for community-oriented research projects (Dijkstra et al., 2017; Dijkstra & Schuijf, 2017).

Public and stakeholder engagement in South Africa is seen as science communication rather than a deliberative model in which stakeholders or the public have a say in the direction of research. The challenge is to find ways to assess and record the impacts of research and innovation on citizens, society or the environment. This could provide a constructive space for transdisciplinary research with social scientists.

Second phase: Implementing RRI in South Africa

In the second phase of the NUCLEUS project, the outcomes of the field trips, the European survey study and the cultural adaptation study were translated into recommendations which provided the basis for a roadmap to guide the implementation of RRI in practice. Ten research institutes and universities served as places where it was possible to experiment with the implementation of RRI. Alongside the sites based in Europe and China, one was situated in South Africa. In this section, experiences from the South African Institute for Aquatic Biodiversity (SAIAB) are described.

As the South African Nucleus Consortium partner, SAASTA was tasked with finding a suitable South African academic institution which could be one of the case studies for implementation of RRI. As a National Research Facility in the NRF, SAIAB was identified as a suitable test site for RRI. Dr Angus Paterson, managing director at SAIAB, and Penny Haworth, SAIAB's manager of communications and governance, were approached to champion the project at SAIAB. The Institute was brought into the project in August 2017. The immediate task was to conduct an RRI self-assessment and develop plans for implementation. For more context on SAIAB, see Box 1.

BOX 1 Setting the scene: SAIAB in South Africa's National System of Innovation

Situated in Grahamstown, recently renamed Makhanda, in the rural Eastern Cape province of South Africa, SAIAB has built on a legacy of ichthyological discovery that began with the ground-breaking discovery of the 'living' coelacanth in 1938. Established as a research institute in 1968, SAIAB is an internationally recognised centre for the study of aquatic biodiversity and in 1999 became a Research Facility of the National Research Foundation. SAIAB is also an Associated Institute of Rhodes University.

Throughout its 50-year history, SAIAB has shown itself to be a consistently transformative space. The institute's origins lie in its long association with the story of the coelacanth through the discovery of this enigmatic prehistoric fish in the nets of a fishing trawler on 24 December 1938 by Marjorie Courtenay-Latimer and its subsequent identification by Rhodes University professor of chemistry, JLB Smith, in early 1939, thus debunking the long-held belief in scientific circles of its extinction.

This was a momentous discovery and the popular media followed the story from the first. Press clippings from 1939 show how the discovery of a 'living fossil' caught the world's imagination. The interest continued as Smith looked in vain for a second, complete, specimen. Finally, in 1952, it was through an advertisement circulated in the media and posters distributed by colleagues and acquaintances, that the elusive second specimen was discovered in the French Comoros islands and brought back to South Africa. This was a moment of special significance for Smith, but it was not his only contribution to aquatic biodiversity science. He and his research partner and wife, Margaret, made numerous expeditions across Africa working on both freshwater and marine fishes; the collections from these expeditions became the core of what is now the National Fish Collection housed at SAIAB.

Smith was a composite communicator and wrote for the popular media as well as scientific journals. His books about the coelacanth story have been translated into numerous languages and his efforts and those of his widow, Professor Margaret Smith, who after Smith's death, founded the JLB Smith Institute of Ichthyology in 1968 and became its first director, created the foundations of ichthyological research in South Africa.

That legacy continued. Scientists intent on tracking down live coelacanths off the coast of Africa received reports of coelacanths being caught by fishermen off the coast of Tanzania and Professor Mike Bruton, who became second director of the JLB Smith Institute after Margaret Smith was instrumental in continuing the research and adding further study specimens to the institute's collection. Bruton has since devoted himself to keeping the story alive through popular publications such as *The Amazing Coelacanth* written for a younger audience and *The Annotated Old Four Legs* which brings the original text of Smith's book, *Old Four Legs* up to date.

When coelacanths were sighted by deep-water divers in Jesser Canyon, Sodwana Bay, off the coast of KwaZulu-Natal in 2000, popular interest in the coelacanth was reignited and the Institute was catapulted to the forefront of marine ecosystems research through the establishment by the DST of a national flagship marine research programme, the African Coelacanth Ecosystem Programme (ACEP). ACEP is managed by SAIAB and is the primary nationally-funded marine research infrastructure programme in South Africa. Under the management of SAIAB's current director, Dr Angus Paterson, it has played an increasingly significant role in the provision of marine research infrastructure to South African universities which otherwise would not have access to such equipment. ACEP has coastal research vessels and equipment based in Durban and Port Elizabeth, and will extend into the Western Cape during 2019.

Through these platforms, SAIAB runs an established marine science transformation programme which provides specialist equipment and training to equip the next generation of scientists and managers with tools to understand and manage environmental change. The ACEP Phuhlisa (Development) Programme is a focused transformation programme which wholly embraces the principles of RRI. Initiated in 2012, it has facilitated access to student bursaries, academic support and equipment to an increasing number of students and their supervisors from historically disadvantaged universities in South Africa. Currently 100 postgraduate students from honours through to PhD are supported at four South African universities - University of Fort Hare, Walter Sisulu University, the University of the Western

Cape, and the University of Zululand.

Research platform provision extends beyond the marine environment. Inland fisheries are highly relevant in southern Africa because they provide an opportunity for socio-economic benefits including jobs, rural livelihoods, food security and economic development based on the small-scale fishing and recreational fishing value chains. Built on significant foundations of taxonomy and systematics in freshwater fishes and freshwater ecology developed under Professor Paul Skelton, SAIAB's third director, SAIAB holds the DST/NRF South African Research Chair in Inland Fisheries and Freshwater Ecology, the overall goal of which is to develop regional capacity and research on inland fisheries to support their sustainable development.

A changing science paradigm

In 1999, the JLB Smith Institute of Ichthyology became a research facility of the National Research Foundation (NRF) which itself had been constituted through an Act of parliament. The science landscape in South Africa was rapidly evolving. The National System of Innovation, a concept promoted by the 1996 White Paper for Science, Engineering and Technology, was facing pressures, challenges and change.

In the following 10 years, the NRF developed rapidly and in 2009 adopted a new strategic plan. In this connection, all national facilities were under scrutiny in terms of their place and role in the National System of Innovation. The PhD epitomised the postgraduate training role that national facility researchers were to embrace. In this context, the facilitating and service-orientated role of national facilities and the importance of flagship programmes that embraced the research community became essential components to consider within the National System of Innovation. However, added to the mix in 2009 there were two overarching components of the research enterprise which applied to all research activities, namely, the need to use research to educate and train students and researchers, and the imperative to link research activities and emerging knowledge to public awareness and information. How SAIAB addressed these components is discussed in more detail in the main body of this chapter.

In 2009, SAIAB was in its 10th year as a national facility. It had established a firm platform for scientific research in aquatic biodiversity in Africa. Furthermore, through various large multi- and inter-disciplinary projects and programmes it had become an effective 'hub' for aquatic biodiversity in southern Africa. Drivers influencing SAIAB's strategic planning over the next ten years included the Biodiversity Crisis - a global concern and obligation; the DST's and Technology Grand Challenges; NRF Vision 2015; SAEON and Long-Term Data sets; the National Environmental Management Act (1998) and its various sub-components such as the Biodiversity Act (NEM:BA 2004); and South African National Biodiversity Institute (SANBI).

Today, SAIAB serves as a major scientific resource for understanding globally significant aquatic ecosystems and has established multi-institutional, multidisciplinary stakeholder networks. SAIAB's research platforms have grown considerably and its scientific leadership and expertise in marine and freshwater aquatic biodiversity are vital to the national interest when dealing with issues arising from exponentially increasing pressures of human population growth and development, climate and global change.

Transformation and social justice at SAIAB

SAIAB had already been through a self-assessment process as part of an institutional review in 2015 and, in preparation for strategic planning towards 2025, had held a workshop with all staff to discuss the institute's vision and mission statement (last revised in 2010) and how this should change to better reflect the institute in 2018 and beyond. However, the RRI self-assessment undertaken for the NUCLEUS project sharpened the focus. These initial steps were of immediate benefit in that they required critical assessment.

Although specific RRI terminology had not been part of the vocabulary of the institute, it was encouraging to find a reasonably well-established culture of RRI. However, the NRF had outlined a draft transformation framework in 2015 and initiated a process of examining diversity in the workplace through professionally facilitated workshops across the NRF in 2015. Workshops run at SAIAB in September of that year, showed that there were still long-standing under-currents of perceived inequity in SAIAB that had to be addressed and that a journey of self-discovery, involving everyone and taking a really good and honest look at the institute, needed to be maintained.

Amid controversial national conversations about transformation, the institute's leadership was not afraid to encourage robust conversations about transformation and diversity. SAIAB's executive showed its commitment to the transformation agenda. In October 2017, the process begun by the NRF continued and a leadership workshop was organised for identified leaders from all job levels at the institute. SAIAB leadership reiterated its determination to work with everyone to address meaningfully the transformation agenda and in April 2018 an all-day, externally facilitated workshop for all staff, interns and students followed. Facilitated by the Wits University Centre for Diversity Studies, which had facilitated the initial series of interactions, the workshop provided an open platform to explore and unpack received and assumed notions of diversity in the workplace. This and the two previous workshops provided stepping-stones for the implementation of transformation at the institute.

After the diversity workshop in April 2018, a workplace transformation committee comprising members identified from all levels of the institution was established to drive the process. The committee instituted some quick wins to appeal to the hearts and minds of those working there, and worked closely with SAIAB's Wellness Committee to offer staff and students access to support mechanisms and activities designed to encourage a positive work-life balance. That said, it is recognised that SAIAB is on a continuous journey in relation to addressing remaining challenges around ethnicity and gender. With this in mind, SAIAB has developed a Transformation and Social Justice Strategy to be integrated into its broader institutional strategy and research agenda for 2019 to 2025.

Reaching out and finding common ground

As mentioned earlier, the intention of RRI is to involve all societal actors in the process of aligning research and innovation outcomes to the values, needs and expectations of society. One of the challenges in the dynamic socio-political context of South Africa that SAIAB had already acknowledged as requiring attention before it became involved in the NUCLEUS project was the need to recognise and optimise relevant science–society links, integrate these into the institute's research strategy and better articulate them to policy-makers and the public. This was addressed, for example, in 2009 at the second Africa Science Communication Conference organised by SAASTA (Haworth, 2009).

As the example below describes, researchers from SAIAB have long recognised their responsibility to contribute to awareness and political action with regard to sustaining biological diversity. Healthy ecosystems and biodiversity are integral to human well-being and sustainable development. However, biological diversity is being lost at an alarming rate due to multiple human impacts, and freshwater fishes and amphibians are ranked among the groups with the highest proportion of species threatened with extinction. Responsibility for future generations requires that co-operative and innovative decisions must be taken now to halt this current trend. In South Africa, the rapid loss of biodiversity is compounded by our incomplete knowledge of species diversity and their geographical distributions.

The latest IUCN Red List assessment of all freshwater fishes of South Africa was done in 2016 by experts from multiple research institutions and conservation agencies. The assessments are produced to communicate findings to policy-makers, environmental managers and the public. SAIAB's legacy of natural history collection management and curation is internationally recognised and provides the platform upon which the institute was brought into the NRF 20 years ago. In 2013, a national survey of natural history collections across South Africa found that SAIAB's National Fish Collection and the associated diversification of the SAIAB collections as a whole to include amphibians, cephalopods, diatoms, tunicates and aquatic invertebrates, were at the forefront of collection curation in South Africa. With this strong foundation in taxonomic research, SAIAB scientists have made significant contributions to IUCN Red List assessments, the most recent of which were published in 2017 and 2018. The data generated has contributed, and will continue to contribute, towards fulfilment of national policy on biodiversity conservation enshrined in the National Environmental Management and Biodiversity Act and fulfilment of the Convention for Biological Diversity's Global Taxonomic Initiative (Chakona et al., 2018).[4]

Box 2 describes a project in which SAIAB's researchers are addressing biodiversity issues through inter-agency collaboration.

BOX 2 Developing a participatory approach to addressing biodiversity issues for human well-being

Evidence from previous and ongoing molecular studies shows that a remarkable proportion of the diversity of freshwater fishes and frogs in South Africa remains scientifically undocumented (Chakona et al., 2015, 2018). The underestimation of taxonomic diversity has profound conservation implications for these threatened groups. In response to an urgent need for an innovative approach that can be used to assign specimens

4 https://www.saiab.ac.za/uploads/files/chakona_saiab_featured_research_november_2018_web_version.pdf

to known species as well as accelerate the pace of species discovery in order to identify priorities for taxonomic research and conservation actions, SAIAB initiated the Topotypes Project in 2014, led by SAIAB senior scientist, Dr Albert Chakona.

Working with regional and international partners, a trans-disciplinary research team comprising postgraduate students, DST-NRF interns and representatives from conservation authorities from the Western Cape (CapeNature), KwaZulu-Natal (Ezemvelo KZN Wildlife) and Mpumalanga (Mpumalanga Tourism and Parks Agency) conducted surveys throughout all 9 provinces in South Africa and collected comprehensive tissue samples and voucher specimens that have been deposited into the National Fish and Frog Collections at SAIAB. Peer-reviewed papers published since the inception of the project include the description of two new species (Chakona et al., 2014; Chakona & Skelton, 2017).

Future research projects will include regional conservation agencies from inception. This will increase appreciation by conservation authorities of the need to include all levels of diversity (species and genetic lineages) in conservation planning. This is integral to the Rio Convention on Biological Diversity and will contribute towards the achievement of the UN's Sustainable Development Goals linked to biodiversity sustainability and provision of ecosystem services to safe-guard human well-being, particularly for impoverished rural communities that are directly dependent on natural resources such as inland fisheries for their survival. Building on this, the challenge will be to ensure that in response to the results, management decisions and the implementation thereof include local communities who are affected by them. Conservation authorities such as CapeNature and Ezemvelo KZN Wildlife have well-developed public communication and environmental education programmes that could provide a vehicle for this.

Identifying and working to existing strengths

In response to the imperative to link research activities and emerging knowledge to public awareness and information as formulated in the NRF Strategy 2009–2014, SAIAB had begun looking for ways to develop a closer working relationship between scientists and communication. The most accessible model for SAIAB to work from in designing its science advancement activities were those it had developed historically and used prior to falling under the NRF. SAIAB was previously a declared cultural institution under Act 29 of 1969, and the engagement model it used was based on museum education and outreach.

For some time, SAIAB had run highly successful education outreach activities, mostly undertaken by dedicated education officers based in the communications division which was not effectively aligned with research activities. To be able to link the institute's research activities and emerging knowledge to public awareness and information, it was essential to find a platform for information transfer between the research division and the communications division. In 2009, the communications manager became included in regular research forum meetings, but the essential character of reported engagement activities remained the established museum education and outreach model. Nevertheless, this was a step in the right direction and the 2009 Africa Science Communication Conference provided an opportunity to share some of the challenges and successes that were experienced in integrating science communication efforts into the strategic imperatives (Haworth, 2009).

Staff changes and shifting priorities in staffing requirements at the institute in 2009 and 2010 resulted in the closure of the education unit at SAIAB. Effectively this put an end to schools outreach. The demise of SAIAB's first education unit and later, through shifting priorities and staff changes, its science communication capacity, meant that SAIAB had to find creative ways in which to try and fulfil its mandate in public engagement as a national facility. Targeted public engagement activities have continued through national focus events such as the DST's National Science Week coordinated by SAASTA and Scifest Africa. Through this shift in available human capacity, researchers, support staff, interns and students have become more involved in formal public engagement activities, but engagement has yet to be fully embedded in the research agenda.

Through shifting priorities and staff changes, SAIAB has lost most of its dedicated science communication capacity. This function has been integrated into the support division and now includes governance as a major component of its focus. To some extent this supports putting RRI in action by finding ways to integrate RRI into the governance structures of academic organisations. However, it has also meant that SAIAB has had to find creative ways in which to try and fulfil its mandate in science communication and engagement.

The first self-evaluation exercise and SWOT analysis undertaken when SAIAB was brought into the NUCLEUS project was conducted from September to October 2017. It showed that SAIAB was implementing aspects of RRI through its management strategies and some of its research projects and related activities although these were not being articulated under that banner. One of the strengths identified in the analysis was SAIAB's position as a long-standing associated institute of Rhodes University. SAIAB's senior scientists are Rhodes University faculty members and SAIAB is represented on the Science Faculty Community Engagement Committee with which it collaborates on events such as faculty open days, Water World at Scifest Africa and National Science Week.

The university places strong emphasis on social innovation and community engagement.[5] The principles underpinning the university's engagement with the Makhanda/Grahamstown community clearly resonate with the principles of RRI, through its stated mission to 'oversee the institutionalisation of community engagement at Rhodes University through the processes of making the

5 https://www.ru.ac.za/communityengagement/about/

university more responsive to its social context and making the university more accessible to the community'.[6] The NUCLEUS project provided a platform through which to find common ground specific to RRI and collaboration began with the SA Research Chair in Biotechnology Innovation and Engagement held by Professor Janice Limson at the Rhodes University Biotechnology Innovation Centre (RUBIC).[7]

In exploring ways to catalyse ongoing debates about the role of science in society, a number of joint activities have been offered. These included a science engagement and dialogue workshop led by Dr Heather Rea from Edinburgh University (2018), combined workshops at Scifest Africa (2018 and 2019) and the National Arts Festival (2018). Experiences in implementing RRI have been presented at the SciComm100 Conference 2018 in Stellenbosch (Haworth, 2018; Limson, 2018). Visits to the Department of Chemistry and Forensics at the School of Science & Technology, Nottingham-Trent University[8] and the steering committee of the Nottingham City Festival of Science and Curiosity,[9] also provided opportunities to share experiences and exchange experiences about RRI within the NUCLEUS project.

Discussion and conclusion: Lessons learned and moving forward

In this chapter we have presented findings from the various studies and experiences conducted as part of the NUCLEUS project which aimed to bring RRI into practice, amongst others, in South Africa. The field trip to South Africa, at the start of the project, showed enthusiasm for engagement among interviewees, although at the same time budget challenges were real. Diversity

6 Rhodes University mission statement, https://www.ru.ac.za/introducingrhodes/visionand-mission/

7 https://www.ru.ac.za/biotech/

8 http://www.nucleus-project.eu/2018/07/26/learn-how-nottingham-trent-university-implemented-rri-in-this-first-phase/

9 http://nottsfosac.co.uk/

and access to education were seen as further challenges to effective engagement. A valid question in such a context is how it is possible to raise science–society relationships to a next level.

The cultural adaptation study provided more insights into RRI elements at the governmental level where innovation is seen as a means to advance the economy and lives of people. Engagement has a place in this. Equal access and inclusion are considered challenges to the science system while there are also examples, such as that of the *kougoed* plant, which show the benefits of collaboration between science and society. A consideration in this regard is that the effects of engagement are hard to measure.

Looking at RRI at the institutional level, it was shown that SAASTA plays an important role in fostering science education and outreach but, despite commendable efforts, rural areas are hard to reach. Institutes embrace open access policies and, according to the interviewees, impact and engagement are considered important, but they are not yet part of the key performance indicators for researchers and therefore their uptake is limited. According to the researchers interviewed, social and environmental impacts of their research should be valued.

Implementing RRI values at the individual level means that many tasks have to be conducted with limited means, which implies, for example, that despite being important, engagement is seen merely as part of the job. Interviewees suggest that role models may stimulate researchers to put effort into RRI-linked aspects. They also believe that careful strategies should be developed to foster inclusion and equality. Engagement is seen as valuable.

In all, elements of RRI were brought into practice at various levels. Although it was often not yet labelled as such, RRI was found in action in many places. While efforts to promote science education and equity are most developed, governance, open access, public engagement and ethics are aspects that needed work.

In the final phase of the NUCLEUS project, and with the findings from the first phase of the project in mind, elements of RRI were put into practice at the research institute SAIAB. Although SAIAB had a long history in engagement and science education, RRI in action sharpened the focus for SAIAB. For example, the conversation about diversity and transformation in dealing with inequality, which had started before the NUCLEUS project, is seen as a continuous journey. Also, the need to optimise and recognise relevant science–society links is better acknowledged. At the same time, examples of specific projects show SAIAB's willingness and ability to do this. In addition, budget constraints cut science education activities but, at the same time, the principles of RRI were incorporated into the governance of the institute. It is also recognised that winning hearts and minds, developing trust and stimulating co-responsibility among all actors at the institutional level is worthwhile. Moreover, at SAIAB, it is clear that RRI elements were already applied but not previously articulated as such. The NUCLEUS project, therefore, has served as a platform to stimulate and find common ground. This is visible, for example, in the newly established collaboration with Professor Janice Limson's DST/NRF South African Research Chair in Biotechnology Innovation and Engagement at Rhodes University, and it is further explicated in various activities and workshops that have been organised, in presentations at conferences, and in sharing experiences.

In other words, at SAIAB there is now more emphasis on catalysing ongoing debates. Lessons for SAIAB, therefore, are that involvement in the NUCLEUS project enabled it to explore a broader context for RRI. It has also consolidated what has up to now been a somewhat fragmented communications and governance portfolio positioned within its support services unit. Furthermore, it has allowed SAIAB to recognise and build on strengths and, importantly, to share this journey with SAASTA which is the NUCLEUS consortium partner in South Africa, and with other (societal) stakeholders in South Africa and abroad.

SAIAB's short-term goal for 2019–2021 is to further embed the principles of RRI within the culture and governance of the institution and the NRF. The immediate challenge is to sustain and continuously build on RRI in action.

Regarding RRI aspects in the broader South African context,

the past few years show an increasing awareness of developing a responsible science–society relationship, which can be further stimulated in the years to come, as pointed out in the White Paper (2019). Translating aspects of RRI aspects in the South African context may require looking at the local level where those aspects can be addressed. For example, our findings show that there is a willingness to engage societal actors in the development of science for meeting societal needs, despite the multiple challenges that exist. As addressed by Cochrane et al. (2019: 6), 'ensuring societal relevance of science and research will bring about benefits but must be accompanied by an increase in expenditure on actionable research and development'.

Finally, some limitations have to be addressed. Findings presented in this chapter are only qualitative in nature and therefore cannot be considered conclusive. Further research and monitoring of RRI – and more specifically the role of engagement and science communication – will be valuable. To this end it is noted that as a follow-up to the NUCLEUS project, SAASTA has collaborated as the representative for Africa with 22 other partners on a three-year project (2018–2021) to form a network of all global RRI projects. The Responsible Research and Innovation Networked Globally (RRING) project takes a bottom-up approach, learning from best practices in RRI globally and from linkages, via the new RRING community, to develop the RRI linked-up world. Its objectives include creating the global RRING community; developing a global open access knowledge base of RRI; aligning RRI to the UN Sustainable Development Goals; determining qualitatively and quantitatively the competitive advantages of RRI; creating high-level RRI strategy recommendations; trialling RRI best practice learning and reviewing EU RRI benchmarking from a global perspective; promoting inclusive engagement of civil society and researchers with the RRING community and open access RRI knowledge base; and to gain social inclusion, co-creation, social innovation and entrepreneurship.[10]

10 http://www.rring.eu/summary/#

To conclude, findings from the NUCLEUS cultural adaptation study in South Africa and the subsequent site implementation at SAIAB proved to be insightful and provide more understanding of what a responsible science–society relationship in the South African context may entail. Further efforts to expand RRI across other NRF research institutes through adopting a similar way of working as described in this chapter, with an RRI mentor working closely with other institute staff to build local capacity, are being considered.

Acknowledgements

This chapter is based on work from the NUCLEUS project, funded by European Union in the Horizon 2020 programme (grant agreement number 664932).[11] We are thankful to Shadrack Mkansi, Mirjam Schuijff and Lin Yin, with whom the case studies in South Africa and China were carried out. We are also thankful to Professor Peter Weingart and Dr Marina Joubert who invited us to the SCICOM100 conference at Stellenbosch University in 2018 where we could present both our results which led to this collaborative chapter.

References

Böger, E. (2017). *Survey conduct report. Deliverable 3.2 NUCLEUS.* http://www.nucleus-project.eu/wp-content/uploads/2016/08/D3_2_NUCLEUS_SurveyConductReport_2017.pdf.

Burget, M., Bardone, E. & Pedaste, M. (2017). Definitions and conceptual dimensions of responsible research and innovation: A literature review. *Science and Engineering Ethics,* 23(1), 1–19. doi:10.1007/s11948-016-9782-1.

Chakona, A., Kadye, W. T., Bere, T., Mazungula, D. N. & Vreven, E. (2018). Evidence of hidden diversity and taxonomic conflicts in five stream fishes from the Eastern Zimbabwe Highlands freshwater ecoregion. *Zookeys,* 768, 69–95. doi:10.3897/zookeys.768.21944.

11 http://www.nucleus-project.eu/

Chakona, A., Malherbe, W. S., Gouws, G. & Swartz, E. R. (2015). Deep genetic divergence between geographically isolated populations of the goldie barb (*Barbus pallidus*) in South Africa: potential taxonomic and conservation implications. *African Zoology,* 50(1), 5–10. doi:10.1080/15627020.2015.1021164.

Chakona, A. & Skelton, P. H. (2017). A review of the *Pseudobarbus afer* (Peters, 1864) species complex (Teleostei, Cyprinidae) in the eastern Cape Fold Ecoregion of South Africa. *Zookeys,* 657, 109–140. doi:10.3897/zookeys.657.11076.

Chakona, A., Swartz, E. R. & Skelton, P. H. (2014). A new species of redfin (Teleostei, Cyprinidae, *Pseudobarbus)* from the Verlorenvlei River system, South Africa. *Zookeys,* 453, 121–137. doi:10.3897/zookeys.453.8072.

Cochrane, K. L., Sauer, W. H. & Aswani, S. (2019). Science in the service of society: Is marine and coastal science addressing South Africa's needs? *South African Journal of Science,* 115(1–2), 1–7.

Commoner, B. (1966). *Science and Survival.* London: Victor Gollancz.

Dijkstra, A. (coordination) with contributions from Dublin City University, Rhine-Waal University, Science View, Aberdeen University, Edinburgh University (2017, 1 November). *NUCLEUS Implementation Roadmap. Deliverable 3.6.* https://ris.utwente.nl/ws/portalfiles/portal/107292320/D3.6_Implementation_Roadmap_.pdf.

Dijkstra, A. M. & Schuijff, M. (2017, 30 March–1 April). Comparison between South Africa and Europe in matters of RRI: Findings from an international case study. Paper presented at the ZiF Conference Responsible Research and Innovation: Coming to Grips with a Contentious Concept, ZiF Bielefeld University, Bielefeld.

Dijkstra, A. M., Schuijff, M., Yin, L. & Mkansi, S. (2017). *RRI in China and South Africa: Cultural adaptation report* (Deliverable 3.3 NUCLEUS project). Enschede: University of Twente. https://issuu.com/nucleusrri/docs/d3.3_nucleus_cultural_adaptation_re.

Doran, H. (2016). *NUCLEUS Field Trip Report: Civil society (Pretoria).* Aberdeen: University of Aberdeen. http://www.nucleus-project.eu/wp-content/uploads/2016/08/D4.4_NUCLEUS_Pretoria-Fieldtrip-Report.pdf

DST (Department of Science and Technology) (2019). *White Paper on Science, Technology and Innovation. Science, technology and innovation enabling inclusive and sustainable South African development in a changing world.* Pretoria: Department of Science and Technology. https://www.dst.gov.za/images/2019/DRAFT_WHITE_PAPER_low_resC.pdf.

Haworth, P. (2009) Integrating public awareness of science and the promotion of science in society into a National Research Facility Science Plan. Paper presented at the 2nd Africa Science Communications Conference 2009: Shaping Africa's future – Science Communication's contribution to Science, Technology and Innovation, and the Development of Democracy in Africa, 18–21 February, Gallagher Convention Centre, Gauteng, South Africa.

Haworth, P. (2018). Beyond science awareness: Embedding responsible research and innovation (RRI) in the research agenda. Paper presented at the SCICOM100 Conference 2018: Science communication and democratic South Africa:

Prospects and Challenges, 5–7 November, Stellenbosch University, South Africa.
Limson, J. (2018). Biotechnology and RRI: exploring models at the science and society interface. Paper presented at the SCICOM100 Conference 2018. Science communication and democratic South Africa: Prospects and Challenges, 5–7 November, Stellenbosch University, South Africa.
Patton, M. Q. (2002). *Qualitative Research and Evaluation Methods* (3rd edn). Thousand Oaks: Sage.
Rip, A. (2014). The past and future of RRI. *Life Sciences, Society and Policy,* 10(1), 17. doi:10.1186/s40504-014-0017-4.
Shelley-Egan, C., Bowman, D. M. & Robinson, D. K. R. (2018). Devices of responsibility: Over a decade of responsible research and innovation initiatives for nanotechnologies. *Science and Engineering Ethics,* 24(6), 1719–1746. doi:10.1007/s11948-017-9978-z.
Unesco (n.d.). South Africa. Global Open Access Portal. http://www.unesco.org/new/en/communication-and-information/portals-and-platforms/goap/access-by-region/africa/south-africa/
Von Schomberg, R. (2013). A vision of responsible research and innovation. In R. Owen, J. Bessant & M. Heintz (eds), *Responsible Innovation: Managing the responsible emergence of science and innovation in society* (pp. 51–74). Chichester: John Wiley.

4 Developing a targeted behavioural change communication strategy for a linguistically and culturally diverse community

Konosoang Sobane & Wilfred Lunga

Introduction

Social and behavioural change communication (SBCC) as a domain in health communication is increasingly being recognised as a valuable tool for modifying lifestyles which pose a threat to people's well-being, and for facilitating improvements in health (Canavati et al., 2016; Hodinot et al., 2017). It becomes particularly valuable in low-middle income countries such as South Africa, where the burden of disease, particularly HIV/ Aids continues to be concerning as noted in recent health surveys such as the South African National HIV Survey V and the South African National Health and Nutrition Examination Survey.

Effective management of disease requires that a patient be adequately informed not only about the nature of the health condition but also about lifestyle and behavioural changes that are commensurate with managing the condition. SBCC initiatives therefore play a vital role in facilitating patient information, thus being one of the enabling factors for patient lifestyle and behavioural changes. Although there has been a plethora of SBCC initiatives, especially for HIV/Aids messaging in South Africa, there has been minimal awareness, among communication developers of the value of involving target communities in the development of these initiatives. There are even fewer documented efforts to base these initiatives on sound theoretical grounding and empirical research that document the needs of the target communities.

This chapter highlights the value of community participation in developing a health communication strategy, and the value of sound theoretical grounding. It unpacks some of the pertinent theories that can inform a health communication strategy. These include the meta-theory of health communication (Kincaid et al., 2012); the health belief model developed by social psychologists the US Public Health Service in the 1950s to explain the relationship between people's health behaviour and perceived risk (Kibler et al., 2018); and theories of communication ecology which originated in the 1970s to describe communication environments and their impact on people (Foth & Hearn, 2007; Scolari, 2012). This chapter also discusses the role of culture in developing and disseminating health messaging (Brincat, 2012), and the importance of choice of communication platform (Felix et al., 2015; Lima et al., 2018) as we argue that these are some of the factors that account for the effectiveness of any SBCC initiative. Lastly, the chapter proposes a framework for the inclusion of the target community in developing a health communication strategy.

The value of using theory in developing communication strategies

Global health organisations have developed strategic foci on health communication, alluding to its significance in promoting health changes in individuals and communities. The Centers for Disease Control and Prevention (CDC) as a leading health organisation is an example. It defines health communication as 'the study and use of communication strategies to inform and influence individual decisions that enhance health' (CDC, 2019: n.p.). This resonates with Schiavio et al.'s (2014: 77) definition of health communication as a 'multifaceted and multidisciplinary field of research, theory, and practice concerned with reaching different populations and groups to exchange health-related information, in order to influence, engage, [and] empower'.

Subsumed in these definitions are multiple ways in which health communication can influence behaviour, namely: (1) creating awareness by providing the target population with information about the health problem, healthcare services and specific actions that people can take to manage or react to the problem (Krige, 2012); (2) improving people's attitudes by emphasising the positive benefits of the behaviour being demonstrated as well as the negative outcomes that may arise if the behaviour is not practiced (see Mutinta, 2012); and (3) connecting and encouraging people to access services by modelling what to expect and how to act when they arrive (see Kunda & Tomaselli, 2009).

These key principles of health communication have made it a potentially useful tool in fighting the concerning prevalence of HIV/Aids, and the continued persistence of non-adherence to antiretroviral medication. Along with this relevance, there has been increasing acknowledgement of the need to ground health communication (and thus SBCC initiatives) on sound theoretical frameworks in order to enhance their effectiveness (Airhihenbuwa & Obregon, 2000). The available theoretical work on health communication is rooted in disciplines such as social psychology, behavioural sciences and communication science. When used as a basis for a strategy, theoretical frameworks may help in predicting the relationship between interventions and behaviour. For example, according to Laranjo (2016), the health belief model posits that there is a relationship between people's likelihood to take preventative action against a health issue and their perceptions of the seriousness of the health threat of that health problem. Grounding a communication intervention in such a theoretical framework therefore helps to unpack how, when and why people would potentially change their behaviour and thus help inform the design of interventions, as well as the appropriate time and context of an intervention. Since they have been grounded on past research, theoretical frameworks can inform the specific actions that a communication intervention can take to influence behaviour changes, and help to predict factors that could potentially hinder or promote these changes. A theoretical framework is therefore a valuable tool in shaping the conceptualisation of an effective communication intervention. The following are some of the pertinent theories that can be drawn on in developing a health communication strategy.

Theories rooted in social psychology

Health communication theoretical frameworks rooted in social psychology draw from health behaviour theory concepts such as stages of change from the trans-theoretical model, self-efficacy from social-cognitive theory, perceived susceptibility from the health belief model, and attitudes, social norms, and behavioural intentions from the theory of reasoned action and planned behaviour (Lee et al., 2015). These concepts are useful in the development of an intervention, particularly a communication one, in that they inform the tailoring that could determine the success of an intervention as noted in Lustria et al. (2013).

The meta-theory of healthcare communication posits that the effectiveness of a health communication intervention is a result of the successful interaction between resources and psycho-social factors that would influence behaviour change (Kincaid et al., 2012). A communication strategy that infuses resources such as promotion, dialogue and advocacy among others, with a set of ideational (psycho-social) factors such as cognitive elements (e.g. beliefs), emotional factors like self-efficacy, and social elements (e.g. interpersonal communication) has a higher potential to influence behaviour change. This potential is increased if the strategy also takes into account other factors such as the socio-economic, political and cultural environment: 'Individuals and their immediate social relationships are dependent on the larger structural and environmental systems: gender, power, culture, community, organisation, political and economic environments' (Manoff Group, 2016: 4). A strategy that has taken into account this comprehensive ecosystem has a potential to influence self-efficacy which, in turn, could positively influence all aspects of human behaviour, including health-related behaviour (Bandura, 2006). When ideational factors that are relevant to understanding a health problem and the need for changed behaviours are addressed, communication programmes are more likely to have a positive impact on health behaviours and ultimately on positive outcomes.

Other social psychology theories tap on the power of role-modelling to change behaviour, and some communication programmes apply these role modelling theories to encourage audiences to model positive behaviours that are presented through the communication intervention, as noted in Bandura (2001). Social cognitive theory (SCT), a cognitive formulation of the social learning theory, is one such theory which asserts that audiences identify with characters who demonstrate behaviour that engages with their emotions, facilitates mental rehearsal and ultimately role-modelling of the new behaviour. SCT, as articulated by Bandura (1986; 2001; 2006) explains human behaviour using a three-way model which presupposes a continuous interaction and reciprocated influence between personal factors, environmental influences and behaviour. The theory is premised on the fact that people learn not only through their own experiences but also by observing the actions of others and the results of those actions. In this way people are more inclined to model characters who demonstrate behaviour that engages with their emotions and ultimately emulate those role models and change into new behaviours (Govender et al., 2013). This role modelling of the new behaviours could ultimately result in encouraging self-efficacy and thus behaviour change (Maloney et al., 2011).

The use of role-model stories is increasingly becoming appropriate for adaptation in the development of health communication interventions. This approach is primarily based on social learning-cognitive theory (Bandura, 1986) wherein role-model stories combine experiences of a 'model' individual in a narrative format that incorporates cultural values, language and local relevancy for targeted communities. Role-model stories share information in 'a non-threatening manner by fostering identification with story characters and experiences, engaging recipients with storyline messages, appealing to personal values and interests, reducing counterarguments against key messages, and improving information retention' (Hinyard & Kreuter, 2006, cited in Berkley-Patton, 2009: 2–3).

Theories informing message design

Other theories inform the process of message design. The communicative ecologies theory is one such theory which asserts that an effective communication strategy needs to be based on evidence of available information resources and practices in a community for which it is intended. The design of a communication strategy can be informed by conducting a communication ecology assessment of the targeted community, or one that is similar to it. The concept of communicative ecology defines a number of mediated and unmediated forms of communication existing in a community (Tacchi et al., 2007).

Foth and Hearn (2007) conceive communicative ecology as having three layers: (1) a technological layer which consists of technologies and connecting media that enable communication and interaction; (2) a discursive layer which is the content of communication available in the community; and (3) a social layer which consists of people and social modes of organising those people. These three layers converge in distinct and localised 'communicative ecologies' (Foth & Hearn, 2007).

The communicative ecology does not ignore the context of the community in terms of who has access to certain resources, power relations and the local economy as well as the socio-economic factors that have a bearing on message access and interpretation. These are all important factors when attempting to understand why certain mediums are used in specific spaces and the personal role that the media plays in people's lives (Tacchi et al., 2007).

The technological layer of communication

In the South African context, the technological layer of communication denotes the mass media (print and broadcast media) and new media technologies (internet and mobile phones) available in communities. The mass media in South Africa, as a commercial enterprise, is generally speaking highly corporate and commercialised. However, for communities where television broadcast is not accessible, community radio plays an important role as an alternative source of information. In rural resource-limited areas community radio is less costly and enables isolated communities to voice their own concerns, while also being informed. For example, ordinary citizens discuss on air issues that are central to them, such as gender relations and combating HIV/Aids, and hence are informative to listeners (Madamombe, 2005).

In terms of new media, Mukund et al. (2010) show that South Africa has one of the largest cell phone coverage in the world, and people use cell phones as a daily communicative tool. Cell-phone containers operated in spaza shops and individual homes are widespread across many townships and rural areas (see Skuse & Cousins, 2008). Because of their accessibility, they are an ideal resource for health communication. Recent statistics on digital population in South Africa show that South Africa has 31.18 million internet users of which 28.99 million are mobile internet users (Statista, 2019). Cell phones are therefore an ideal resource for health communication, and present an opportunity for a wider reach for antiretroviral adherence communication programmes that use cell phones as a communication tool.

The social layer of communication

The social layer of the communicative ecology consisting of community organisations, rallies, community meetings, social clubs (stokvels) and churches provides useful and alternative communicative spaces where social networks are forged and strengthened. These are forms of unmediated communication: face-to-face/interpersonal communication not done through any channel of media, as opposed to mediated communication which is done through different form of media. The social communication spaces play important roles in broader community struggles for social and economic development (Chiumbu, 2010; Wilkinson, 2013). These kinds of social networks that are already available in most South African grassroots communities can be tapped into as platforms for communicating HIV treatment adherence messages. Their value is that they already have strong roots in the community and have insider perspectives of adherence issues in their locality. Such community networks have already been used successfully in other countries such as Malawi (Zachariah et al., 2006) and South Africa (Wilkinson, 2013).

The extended parallel process model (EPPM) positions message design at the centre of potential responses by the target community. The theory posits that if messages are framed as threats, an individual's response involves two distinct cognitive appraisals (Witte, 1992). The first appraisal relates to the degree to which the message is perceived as threatening (i.e. how susceptible an individual believes they are to the threat and how severe the consequences would be should the threat occur). If the individual perceives that they are personally vulnerable and the threat is severe, a second appraisal, coping appraisal, occurs whereby the individual considers whether the message provides effective and useful strategies (i.e. 'response efficacy'), and whether they believe that they possess the ability to enact such strategies (i.e. 'message self-efficacy') to help avoid/reduce the threat (Witte, 1992, 1994). In other words, the extent to which an individual is fearful in response to the message's threat (as a result of the first appraisal), determines whether they are motivated to continue processing the message. In turn, the coping appraisal determines the nature of an individual's response to a message and whether they initiate adaptive (danger control) or maladaptive (fear control) processes which correspond to message acceptance and message rejection respectively (Witte, 1992, 1994). EPPM assigns a more significant role to the emotion of fear than some of the EPPM's theoretical predecessors (Witte, 1992; Witte & Allen, 2000). In the EPPM, if a threat is considered relevant and severe, the emotion of fear is posited to ensure ongoing processing of the message and efficacy will determine whether an individual seeks to control the threat (danger control) or to control the fear (fear control) (Witte, 1992; Witte & Allen, 2000). Thus, the emotion of fear may be considered important for individuals' attention and functioning to ensure ongoing processing.

The effectiveness of any communication strategy therefore requires in-depth understanding of the targeted population. This understanding can only be fully achieved if the intended population is actively involved, providing insider knowledge of the dynamics of their own communities, and being collaborators in what can work in their own context.

The role of culture in shaping peoples' understanding of health messaging in different contexts

As health communication continues to be seen as important and the need for communication contextualisation is increasingly understood, there has evolved a consensus that culture also has to be taken into account in designing messaging. This is in acknowledgement of culture as a factor that can influence health and health behaviours (Tseng, 2001). In-depth understanding of the cultural characteristics and practices of a given group allows communication interventions to be customised to meet the needs of people affected or who are at risk. Spencer-Oatey (2000) conceptualises culture as set of attitudes, beliefs, behavioural conventions and basic assumptions and values that are shared by a group of people, and that influence each member's behaviour and each member's interpretations of the 'meaning' of other people's behaviour. This is consistent with Samovar et al.'s (2012) definition of culture as the rules for living and functioning in a certain society. These rules determine and influence how members of a community generally behave, and, as a consequence, community action and reaction to messaging is oriented by culturally mediated beliefs about what is real and what is good.

The calls for the incorporation of culture in designing health promotion messaging (Airhihenbuwa, 1995) were made in cognizance of the fact that different cultures differ in their descriptions, conceptualisations and experiences of health problems, their causes, perceptions of how to react to the problem. When these differences are not well managed in designing messages, misunderstandings are likely to occur and the health communication strategy is likely to become ineffective in influencing the necessary behaviour changes.

Culturally sensitive and culture-centred health communication approaches

Efforts to ground health communication and health promotion on culture have resulted in two research-based approaches, namely, the cultural sensitivity approach and the culture-centred approach (Dutta, 2007). Betsch et al. (2017) define culture-sensitive health communication as an approach that makes deliberate efforts to engage in evidence-informed adaptation of health communication to the recipients' cultural background with an aim of enhancing the persuasiveness and thus effectiveness of messages in health promotion. The adaptation can be in the form of incorporating culturally appropriate and sensitive terminology to a health communication strategy (Dickerson et al., 2018). This is expected to lead to self-efficacy and improves recipient's preparedness for health medical decision-making. The goal of the cultural sensitivity approach is therefore to ensure that message content and message framing is sensitive to the culture, and that culturally appropriate terminologies and language forms are used. This approach has, however, been criticised for often being superficial and for sometimes tailoring the message content to an already available approach that does not fully respond to the cultural spectrum of a target audience.

While culturally sensitive approaches focus more on adaptation, culture-centred approaches harness culture-specific knowledges of the target communities and employ co-creation and co-development of communication strategies with the communities. Culture-centredness embraces communication strategies that utilise indigenous history, language and values as a basis. In this way the intervention helps the target communities to 'both decolonise and reclaim their cultural beliefs, practices, and aspirations that promote health and well-being' (Dickerson et al., 2018). Adding the principles of co-creation to this culturally-centred knowledge facilitates the sustainability of a health community strategy. This is because there is a sense of inclusion and ownership by the targeted population from the initial stages of development. The approach recognises the value of using community agency, strengths, power and language as a foundation to sustainability and a facilitator of health behavioural changes (Belone et al., 2016, 2017; Dickerson et al., 2018; Dutta, 2007). Basing a communication strategy on the co-creation with a community and on local cultural knowledge, practices and aspirations can improve their ultimate efficacy.

There is a body of literature that sees the two approaches as alternatives (e.g. Dutta, 2007; Okamoto et al., 2014). However, there is also a large body of literature that sees them as a continuum (e.g. Dickerson et al., 2018). We argue that both approaches could yield positive results if they are used to complement one another. Each one affords the process of developing a communication strategy that taps into culture at different spheres to produce a strategy richly informed by in-depth understanding of the target community.

Empirical examples of successful culturally-oriented health communication in the Global South: a specific focus on HIV/Aids

Since HIV/Aids continues to be one of the global health threats, most communication campaigns have targeted this area. Most of the campaigns have developed communication strategies and interventions which often take community-based approaches and capitalise on co-creation with local communities to enhance effectiveness. They therefore provide valuable insights into the value of being aware of, and incorporating culture in developing a communication intervention.

Some of these successful interventions, based on co-creation with local communities, have been reported in countries such as

Thailand (Nelson et al., 1996), Ethiopia and Uganda (Gusdal et al., 2011). In Uganda and Ethiopia, an adherence to an antiretroviral (ART) programme was implemented in both countries for a HIV-infected cohort. Part of the programme was to involve peer counsellors as facilitators of adherence. They acted as role models, and were involved in the development and dissemination of awareness messaging in face-to-face interactions. Their involvement in the process ensured that culturally appropriate messages were developed and disseminated in local languages through culturally sensitive interactive processes. Since they had insider knowledge of the cultural dynamics of the community, their involvement provided patients with an opportunity to individually talk to someone who was also living with HIV, who had a positive and life-affirming attitude about their situation, and were willing to share personal stories of hope when educating and counselling their patients. The programme had very positive effects on ART adherence and its success was attributed to the involvement of community members (Gusdal et al., 2011).

In South Africa, we draw from two examples both of which involved the involvement of local community members to disseminate ART adherence education. Peltzer et al. (2012) reports on the effectiveness of an ART adherence intervention programme implemented in a hospital in KwaZulu-Natal, involving information-sharing group sessions facilitated by a trained lay health-worker from the local community. The involvement of this facilitator brought with it communication in a local language and in-depth knowledge and sensitivity to the culture of the community. Wilkinson (2013) reports on the successes of the adherence club intervention in Khayelitsha, which employed the strategy of using group meetings facilitated by a local facilitator who provided adherence education as part of the intervention package. The common thread in the two interventions is dissemination of health communication through a local facilitator, thereby tapping on the local knowledge of the cultural dynamics.

In Cambodia, Manavati et al. (2016) report on a malaria-related intervention implemented in cognizance of, among others, the rising concerns about the spread of drug resistance. A conflation of intense social and behavioural change communication (SBCC) activities was implemented by different organisations working in areas identified for malaria containment. This intervention package included new SBCC messages to inform and create awareness among the public. The SBCC intervention package was as follows (Canavati et al., 2016):

- Dissemination of media products for broadcasting on radio and television;
- Working with village health volunteers to communicate messages to encourage preventive and positive health-seeking behaviour in communities, build the capacity of health-centre staff and improve the utilisation of the public health system for malaria diagnosis and treatment; and
- Working with village malaria workers and mobile malaria workers, who are community members, to encourage preventive and positive health-seeking behaviour in communities.

In that way, the intervention involved community members to do health communication in their own communities and in local languages. The shared community membership subsumed with it in-depth knowledge of the psycho-social factors that surround the health issue in the community, and the relevant cultural considerations to be made in communicating to these communities. The intervention's successes of positive improvements in both attitudes and behaviours among the population, and the increase in people seeking treatment for fever, can be attributed to the involvement of insiders to the community.

These examples highlight the value of the involvement of people who understand the culture and language of the community in health communication activities. These people fully understand the cultural aspects of the disease and how culture has determined the way in which people in that community describe and react to the disease. They also fully understand the local language and are conversant with cultural sensitivities related to any of these diseases. Their communication is therefore culturally appropriate and they know how to make the content both culturally appropriate and sensitive. An effective health communication strategy should therefore consider the locally available package of knowledge, resources and skills and use these resources.

Choice of communication platform as an enabler for the effectiveness of a health communication strategy

The success of a health communication strategy also depends on the choice of communication platform and the extent to which the platform is accessible to the target community. Given the changing patterns in communication, marked by a shift from traditional ways of communication to digital platforms, there has also been a transition to digital platforms in health communication. In this section we discuss some of these novel platforms in health communication and what makes each of them successful.

MHealth Interventions

The use of mobile phones in health intervention programmes has increasingly gained momentum over the years. These platforms, collectively called 'mHealth', have afforded communication organisations wider reach given the penetration of mobile phones even to remote areas. The World Health Organization (2011) in its report 'mHealth: New horizons for health through mobile technologies' identifies several categories of mHealth such as health call centres, mobile telemedicine, appointment reminders, community mobilisation and health promotion, mobile patient records, information access, patient monitoring, health surveys and data collection, surveillance, health awareness raising, and decision support systems. mHealth, especially in the area of HIV/Aids is growing in Africa. According to Klasnja and Pratt (2012), mobile technology is a particularly attractive tool for delivering health interventions, due to: (1) its widespread adoption and potential for powerful technical capabilities; (2) the tendency to carry mobile phones everywhere; (3) people's emotional attachment to phones; and (4) the context-awareness features of mobile phones that allow for personalisation. In particular, texting is an effective way to educate and support under-served and diverse populations due to its mass reach and relatively low cost (Fjeldsoe et al., 2009). For health purposes, texting can be used to (1) enhance health service provision (i.e. appointment reminders, vaccination reminders); (2) distribute mass health education messages (i.e. disease outbreaks); (3) encourage better disease self-management practices; and (4) deliver personalised health promotion interventions (Fjeldsoe et al., 2012). Studies elsewhere have also found that mobile phones have advantages when used in health programmes for the youth, as young people in general are responsive to and excited about using new technologies.

For a successful mHealth platform in South Africa, we draw on the example of the MomConnect platform, a multi-faceted programme aimed at promoting demand for maternal health services as well as improving the supply and quality of those services to expectant mothers. The programme includes stage-based health messages sent by SMS to expectant mothers, a text-based help desk that provides answers to pressing questions, a library of health information accessed via a USSD menu, among others An evaluation of the programme by Skinner et al. (2018) showed that the target community was happy with the service, citing valuable content, accessible medium, and ability to save messages for further empowering others as the strengths of the platform. In that way the platform was effective and it met the target population's needs.

A second example is the Biolink platform developed for implementation in KwaZulu-Natal with the aim of linking people living with HIV with care (Comulada et al., 2018). Patient home visitation data is stored in an Android smartphone application and is accessible to the research team and to clinic staff. When clinic staff identify someone who is infected, who is not linked to care and who does not collect prescription refills within a specified time, SMS alerts are sent to field staff mobile phones so that they can follow-up with their assigned patient to facilitate linkage and/or medication adherence. Such use of cell phones as a platform was reportedly successful, due to the high cell phone coverage and the cost effectiveness of the platform. This resonates with the observation that SMS is a favoured approach in South Africa and low- and middle-income countries due to its low cost and flexibility across mobile systems relative to other mHealth tools, such as mobile applications (Lester et al., 2010; Mukund & Murray, 2010).

Beyond South Africa, an example of a programme that used SMS messages is the Kenya Weltel programme implemented from 2007 to 2009 to improve adherence to ART (Van der Kop et al., 2012). The intervention involved sending weekly messages to patients inquiring how they were doing, and participants were required to respond either that they were well or that there was a problem. In a randomised controlled trial, Van der Kop et al. (2012) show that weekly text messages led to improved ART adherence and viral load suppression among those initiating ART. The intervention enabled frequent communication between clinicians and patients, and many patients valued the service for the support it provided and for its cost-effectiveness and ease of accessibility.

Social media

Beyond text messaging, the use of social media in health communication interventions is being adopted, although the current body of knowledge on this is limited. Social media is increasingly being considered as an innovative tool in SBCC because of its capacity to target and reach diverse audiences since it is not limited by space or time.

Social media is an inexpensive, effective method for delivering public health messages. Based on the understanding that SBCC is not merely the transmission of health information to passive audiences, the multi-directional interactivity in social media offers an unmatched advantage (Adams, 2010; Taylor, 2012). Not only is social media being used in searching for health information, clients now get involved directly in managing their health conditions through the use of social media (Campbell & Craig, 2014, cited in Adewuyi & Adefemi, 2016). Although social media is increasingly being used by public health departments in developed countries, it is not yet clear how best to capitalise on social media for raising awareness and, ultimately, triggering behavioural change (Gough et al., 2017). In Africa, the use of social media in health communication campaigns is still very low. There is room for expansion despite digital inequality or divide, to creatively use social media for communication campaigns in the HIV sector. Research shows that mobile social media use in Africa is increasing year by year (Gough et al., 2017).

These technology-based communications highlight the need to be innovative and to capitalise on new and accessible technologies when developing a communication strategy. This implies that prior to developing a strategy there is a need to conduct an in-depth needs assessment and a communication ecology to identify platforms that will be acceptable and accessible for the target community. There is also a need to gain insight into what resources are already available and how they can be harnessed for use in the community.

Conclusion

The discussion in this chapter shows that the effectiveness of a health communication strategy is a result of multiple factors all of which entail a deep understanding of the target community by communication developers and implementers. The theories discussed illuminate the value of understanding the cognitive behaviours and decision-making practices of an audience before developing a communication strategy. The theories also underscore the importance of understanding psycho-social behaviours such as attitudes towards a particular health problem, in order to predict potential reactions to messaging. This knowledge is instrumental in informing message content and delivery.

The chapter also highlights the need to understand and incorporate culture in designing a health communication strategy. This entails the deliberate adaptation of messaging to culture and deliberate co-creation and co-development with locals as a resource which has insider knowledge of the cultural dynamics of a given community. When message content has been developed, dissemination also requires tapping into the local community and harnessing their own 'knowledges' in conveying the messages in order to enhance acceptability of the messages and facilitate reaching the desired health behaviour changes.

Ultimately, all these point to a need for the involvement of the target community in all aspects of development and dissemination of health communication strategy. In particular, their involvement should go beyond the channel of communication and delve into their participation as co-producers in order to ensure the sustainability of the strategy. We argue that the development of a health communication strategy should not take the 'us for them' approach where developers own all the processes and use locals for dissemination, but an 'us with them, for them approach' emphasising the principle of co-creation of knowledge.

Framework for community inclusion in developing a strategy

Given the obvious need to involve the target community, we propose the following as a framework for the inclusion of the target population. The framework is an integration of insights from this paper and principles from Netto et al. (2010). The proposed framework involves the following steps:

- Conducting a needs assessment to establish the health communication needs of a community, and identifying the already available communication platforms and current practices. This should be done in collaboration with the target community to get community-relevant information. Using the resources is likely to increase intervention accessibility.

- In developing the strategy, involve the segments of the target population throughout the process to ensure cultural appropriateness of the content a delivery plan. Harness the knowledge they already have and incorporate such knowledge in the strategy. There is also a need to harness the significance of cultural and religious authorities due to the influence they have on communities. The involvement of this cadre of society and the use of already locally recognisable knowledge has the potential of being easily accepted and likely to influence the desired change.
- In choosing the platforms of communication, it is valuable to consider the use of platforms that already have a wide reach in the community. This does not only facilitate positive user experiences, but also facilitates accessibility, which will increase the reach of the health messaging and potentially have a positive impact. Dissemination should also be by locals who speak local languages and are able to respond to questions in the language of the community and in culturally appropriate ways.

With this proposed involvement of the target audience in all facets of the development of a health communication strategy, we argue that there is increased potential for efficacy and sustainability.

References

Adams, R. J. (2010). Improving health outcomes with better patient understanding and education. *Risk Management and Healthcare Policy*, 3, 61.

Adewuyi, E. O. & Adefemi, K. (2016) Behavior change communication using social media: A review. *International Journal of Communication and Health*, 9, 109–116.

Airhihenbuwa, C. (1995). *Health and culture: Beyond the Western paradigm*. Thousand Oaks, CA: Sage.

Airhihenbuwa, C. O. & Obregon, R. (2000). A critical assessment of theories/models used in health communication for HIV/AIDS. *Journal of Health Communication*, 5(sup1).

Bandura, A. (1986). Fearful expectations and avoidance actions as coefficient of perceived self-inefficiency. *American Psychologist*, 41, 1389–1391.

Bandura, A. (2001). Social cognitive theory of mass communication. *Media Psychology*, 3, 265–298.

Bandura, A. (2006). On integrating social cognitive and social diffusion theories. In A. Singhal & J. Dearing (eds), *Communication of Innovations: A journey with Ev Rogers*. Beverly Hills: Sage.

Belone, L., Lucero, J. E., Duran, B., Tafoya, G., Baker, E. A., Chan, D. & Wallerstein, N. (2016). Community-based participatory research conceptual model: Community partner consultation and face validity. *Qualitative Health Research*, 26(1), 117–135.

Belone, L., Orosco, A., Damon, E., Smith-McNeal, W., Rae, R., Sherpa, M. L., et al. (2017). The piloting of a culturally centered American Indian family prevention program: A CBPR partnership between Mescalero Apache and the University of New Mexico. *Public Health Reviews*, 38, 13.

Berkley-Patton, J., Goggin, K., Liston, R. & Bradley-Edwin, A. (2009). Adapting effective narrative-based HIV prevention interventions to increase minorities' engagement in HIV/AIDS services. *Health Communication*, 24:199–209.

Betsch, C., Böhm, R., Korn, L. & Holtmann, C. (2017). On the benefits of explaining herd immunity in vaccine advocacy. *Nature Human Behaviour*, 1(3), 0056.

Brincat, M. (2012). Medication adherence: patient education, communication and behaviour. *Journal of the Malta College of Pharmacy Practice*, 3–5.

Canavati, S. E., et al. (2016). Evaluation of intensified behaviour change communication strategies in an artemisinin resistance setting. *Malaria Journal*, 15(59), 1–14.

Centres for Disease Control and Prevention (CDC) (2019, 12 August). Health communication basics. CDC website. https://www.cdc.gov/healthcommunication/healthbasics/WhatIsHC.html

Chiumbu, S. (2016). Media, race and capital: A decolonial analysis of representation of miners' strikes in South Africa. *African Studies*, 75(3), 417–435.

Dickerson, D., et al. (2018). Encompassing cultural contexts within scientific research methodologies in the development of health promotion interventions. *Prevention Science*, 1–10.

Dutta, Mohan J. (2007). Communicating about culture and health: Theorizing culture⊠centred and cultural sensitivity approaches. *Communication Theory*, 17(3), 304–328.

Felix, R., Rauschnabel, P. & Hinsch, C. (2017). Elements of strategic social media marketing: A holistic framework. *Journal of Business Research*, 70, 118–126.

Fjeldsoe, B., Miller, Y. & Marshall, A. (2012). Text messaging interventions for chronic disease management and health promotion. In S. Noar & N. Harrington (eds), *E-Health Applications: Promising Strategies for Behavior Change* (pp. 167–186). New York, NY: Routledge.

Fjeldsoe, B. S., Marshall, A. L. & Miller, Y. D. (2009). Behavior change interventions delivered by mobile telephone short-message service. *Am J Prev Med*, 36, 165–173.

Foth, M. & Hearn, G. (2007). Networked individualism of urban residents: discovering the communicative ecology in inner-city apartment buildings. *Information, Communication & Society*, 10(5), 749–772.

Gough A., Hunter R.F. & Ajao, O. (2017). Tweet for behavior change: using social media for the dissemination of public health messages. *JMIR Public Health Surveillance*, 3, e14.

Govender, N. P., Meintjes, G., Bicanic, T., Dawood, H., Harrison, T. S., Jarvis, J. N., et al. (2013). Guideline for the prevention, diagnosis and management of cryptococcal meningitis among HIV-infected persons: 2013 update. *South African Journal of HIV Medicine,* 14, 76–86.

Graziose, M. M., Downs, S. M., O'Brien, Q. & Fanzo, J. (2012). Systematic review of the design, implementation and effectiveness of mass media and nutrition education interventions for infant and young child feeding. *Public Health Nutrition*, 21(2), 273–287.

Gusdal, A. K., Obua, C., Andualem, T., Wahlstrom, R., Chalker, J. & Fochsen, G. (2011). Peer counselors' role in supporting patients' adherence to ART in Ethiopia and Uganda. *AIDS Care,* 23(6), 657–662.

Gusdal, A. K., Josefsson, K., Thors Adolfsson, E. & Martin, L. (2016). Informal caregivers' experiences and needs when caring for a relative with heart failure: An interview study. *Journal of Cardiovascular Nursing,* 31, 1–8.

Hodinot, J., Ahmed, I., Ahmed, A. & Roy, S. (2013). Behaviour change communication activities improve infant and young child nutrition knowledge and practice of neighbouring non participants in a cluster-randomised trial in rural Bangladesh. *PLoS One*, 12(6), e0179866. doi: 10.1371/journal.0179866

Kibler, J. L., Ma, M., Hrzich, J. & Roas, R. (2018). Public knowledge of cardiovascular risk numbers: Contextual factors affecting knowledge and health behavior, and the impact of public health campaigns. In R. R. Watson & S. Zibadi (eds), *Lifestyle in Heart Health and Disease*. London: Academic Press.

Kincaid, D. L., Delate, R., Figueroa, M. E. & Storey, D. (2012). Advances in theory-driven design and evaluation of communication campaigns: Closing the gaps in practice and in theory. In R. Rice & C. Atkin (eds), *Public Communication Campaigns* (4th edn, pp. 305–319). Thousand Oaks, CA: Sage.

Klasnja, P. & Pratt, W. (2012). Healthcare in the pocket: Mapping the space of mobile-phone health interventions. *Journal of Biomedical Informatics*, 45, 184–198.

Krige, D. (2012). Health communication: The case of TB information leaflets. In K.G. Tomaselli & C. Chasi (eds), *Development and Public Health Communication* (pp. 228–247). Cape Town: Pearson South Africa.

Kunda, J. E. L. & Tomaselli, K. (2009). Social representations of HIV /AIDS in South Africa and Zambia: Lessons for health communication. In L. Lagerwerf, H. Boer & H. Wasserman (eds), *Health Communication in Southern Africa: Engaging with social and cultural diversity*. Amsterdam: Rozenberg.

Lee, J. Y., Park, H. A. & Min, Y. H. (2015). Transtheoretical model-based nursing intervention on lifestyle change: a review focused on intervention delivery methods. *Asian Nursing Research*, 9(2), 158–167.

Laranjo, L., Dunn, A. G., Tong, H. Y., Kocaballi, A. B., Chen, J., Bashir, R. et al. (2016). Conversational agents in healthcare: A systematic review. *Journal of the American Medical Informatics Association*, 25(9), 1248–1258.

Lester, R. T., Ritvo, P. & Mills, E. J. (2010). Effects of a mobile phone short message service on antiretroviral treatment adherence in Kenya (WelTel Kenya1): A randomised trial. *Lancet,* 376, 1838–1845.

Lima, S. G. C., Sousa-Lima, R. S., Tokumara, R. S., Nogueira-Filho, S. L. G. & Nogueira, S. S. G. (2018). Use of the Whatsapp application in health follow-up of people with HIV: A thematic analysis. *SciELO Brasil,* 22(3), 1–6.

Lustria, M. L., Noar, S. M., Cortese, J., Van Stee, S. K., Glueckauf, R. L. & Lee, J. (2013). A meta-analysis of web-delivered tailored health behavior change interventions. *Journal of Health Communication*, 18, 1039–1069.

Madamombe, I. (2005). Community radio, a voice for the poor. *Africa Renewal Magazine.* https://www.un.org/africarenewal/magazine/july-2005/community-radio-voice-poor.

Maloney, E. K., Lapinski, M. K. & Witte, K. (2011). Fear appeals and persuasion: A review and update of the extended parallel process model. *Social and Personality Psychology Compass*, 5(4), 206–219.

Manoff Group (2016) Technical Brief: Defining social and behavioural change communication (SBCC) and other essential health communication terms. http://manoffgroup. com/documents/DefiningSBCC.pdf

McDaniel, B. T., Loyne, S. M. & Holmes, E. K. (2012). New mothers and media use: Association between blogging, social networking and maternal wellbeing. *Maternal Child Health*, 16(7), 1509–1517.

Mukund, Bahadur K.C. & Murray, P. J. (2010). Cell phone short messaging service (SMS) for HIV/AIDS in South Africa: a literature review. *Studies in Health Technology and Informatics*, 160(1), 530.

Mutinta, M. A. (2013). Factors contributing to unsafe sex among teenagers in secondary schools of Botswana. University of Botswana, Unpublished master's dissertation.

Nelson, K. & Cooprider, J. (1996). The contribution of shared knowledge to IS group performance. *MIS Quarterly*, 20(4), 409–432.

Netto, G., Bhopal, R., Lederle, N., Khatoon, J. & Jackson, A. (2010). How can health promotion interventions be adapted for minority ethnic communities? Five principles for guiding the development of behavioural interventions. *Health Promotion International,* 25(2), 248–257.

Okamoto, S. K., Kulis, S., Marsiglia, F. F., Steiker, L. K. H. & Dustman, P. (2014). A continuum of approaches toward developing culturally focused prevention interventions: From adaptation to grounding. *The Journal of Primary Prevention*, 35(2), 103–112.

Peltzer, K., Ramlagan, S., Jones, D., Weiss, S. M., Fomundam, H. & Chanetsa, L. (2012). Efficacy of a lay health worker led group antiretroviral medication adherence training among non-adherent HIV-positive patients in KwaZulu-Natal, South Africa: results from a randomized trial. SAHARA-J: *Journal of Social Aspects of HIV/AIDS*, 9(4), 218–226.

Schiavo, R., May, L. M. & Brown, M. (2014). Communicating risk and promoting disease mitigation measures in epidemics and emerging disease settings. *Pathogens and Global Health*, 108(2), 76–94.

Scolari, C. A. (2012). Media ecology: Exploring the metaphor to expand the theory. *Communication Theory*, 22(2), 204–225.

Skuse, A. & Cousins, T. (2008). Getting connected: the social dynamics of urban telecommunications access and use in Khayelitsha. Cape Town. *New Media & Society*, 10(1), 9–26.

Spencer-Oatey, H. (2000). Introduction: Language, culture and rapport management. In H. Spencer-Oatey (ed.), *Culturally Speaking: Managing rapport through talk across cultures* (pp. 1–8). London: Continuum.

Statista (2019). Digital population in South Africa as of January 2019 (in millions). Statista website. https://www.statista.com/statistics/685134/south-africa-digital-population/

Tacchi, J., Fildes, J., Martin, K., Mulenahalli, K., Baulch, E. & Skuse, A. (2007). Communicative ecology. In: *Ethnographic Action Research Handbook*. Kelvin Grove: Queensland University of Technology. http://ear.findingavoice.org/intro/2-0.html.

Taylor, H. (2012). *Social Media for Social Change: Using the internet to tackle intolerance*. Institute for Strategic Dialogue. http://www.theewc.org/content/download/1892/14891/file/Social_Media_Social_Change%20(2).pdf.

Tseng, P. (2001). Convergence of a block coordinate descent method for non differentiable minimization. *Journal of Optimization Theory & Application*, 109 (3), 474–494.

Van der Kop M. L., Karanja S., Thabane L., Marra C., Chung M.H., Gelmon L., et al. (2012) In-depth analysis of patient-clinician cell phone communication during the WelTel Kenya1 antiretroviral adherence trial. *PLoS One*, 7(9).

Wilkinson, T. M. (2013). Nudging and manipulation. *Political Studies*, 61(1), 341–355.

Witte, K. (1992). Putting the fear back into fear appeals: The extended parallel process model. *Communication Monographs*, 59, 329–349.

Witte, K. (1994). Fear control and danger control: A test of the extended parallel process model (EPPM). *Communication Monographs*, 61, 113–132.

Witte, K. & Allen, M. (2000). A meta-analysis of fear appeals: Implications for effective public health campaigns. *Health Education & Behavior*, 27, 591–615.

World Health Organization (2011). *MHealth: New Horizons for Health Through Mobile Technologies*. Volume 3 of Global Observatory for Ehealth. Geneva: World Health Organization.

Zachariah, R., et al. (2006). How can the community contribute in the fight against HIV/AIDS and tuberculosis? An example from a rural district in Malawi. *Transactions of the Royal Society of Tropical Medicine and Hygiene*, 100(2), 167–175.

5 The challenge of communicating science effectively in fisheries management

Doug S. Butterworth

Introduction

This chapter presents a short case study of a particular field of science communication: scientific advice to decision-makers[1] in fisheries management. Scientific advice to decision-makers is a special type of science communication, as it is directly linked to political decisions. Thus, the advisory process requires certain institutional structures that guarantee the quality of the process.

The chapter commences with a very brief introduction to fisheries management: what are the basic objectives, and what makes them difficult to attain. It then proceeds to summarise the basic structures that underlie the process of developing scientific advice for fisheries management measures, and of transmitting this advice to decision-makers (such as the government ministers responsible) for final decisions. This is discussed both in a South African and international context, addressing whether they are working and where the problems lie.

Finally, problems in the way scientists try to communicate scientific results in these processes are highlighted, with some suggestions given of how they might be improved.

---

1 'Decision-maker' is the term customarily used in fisheries management; other fields may refer to 'policy-maker'.

Fisheries management (in brief)

At base fisheries management has two objectives. The first is the sustainable utilisation of a renewable marine resource such as a fish population – both use for the present, and also maintaining the potential to continue that use indefinitely into the future. The second is the recovery of depleted resources – if previous over-exploitation has depleted a resource to a low level of abundance where it can provide only a small sustainable yield, facilitate that resource's growth back to a higher abundance where that sustainable yield will be larger (ideally to provide the maximum sustainable yield).

Sustainable utilisation mirrors the situation of a pensioner who retires with a lump sum invested in a bank. To continue to live 'sustainably', the pensioner must live off the interest, and not dip into capital. The pensioner's annual budget process is simple: multiply the sum invested by the interest rate per annum offered by the bank, and then ensure that projected annual expenditure does not exceed the result of that computation.

So why is the equivalent process not as straightforward in fisheries management? The answer is that fisheries have uncooperative bank tellers. Typically, the information available for the capital multiplied by interest rate computation has the following features:

1. Capital:
    - The amount is advised only once a year (e.g. by a sampling survey of the resource conducted by a research vessel in what is an expensive exercise).
    - Typically, the result will be in error by some 25%.
    - The units differ from those for amounts withdrawn (the catch taken), e.g. balance in roubles, withdrawals in rands, and no information on the exchange rate.

2. Interest rate:
    - No advice is available on the interest rate.
    - The rate varies greatly from year to year – often within a

range from negative by up to half the overall average to positive and triple that average.
- The value of that average rate has to be inferred from few and noisy data.

Hence the computation each year of a catch level that will be sustainable is much more difficult than the multiplication which a pensioner needs to do. In essence, the underlying problem stems from the fact that fisheries is an inexact science.

Furthermore, difficult trade-off decisions are required. For a depleted resource, rapid recovery will require a large reduction in the current catch; with slower recovery, that reduction can be smaller. Unsurprisingly, commercial interests tend to favour the latter given the immediate losses in profitability and employment associated with the former. Small-scale industry has an even greater preference for the latter option because it will typically not have the access to the cash reserves which larger companies can use to see out a period of poor financial returns before resource improvements (hopefully) occur later. In contrast, environmental non-government organisations (ENGOs), such as the World Wide Fund for Nature (WWF) and Greenpeace, will advocate the rapid recovery option with a longer-term perspective in mind.

Many interest groups and professions can become involved in the resultant discussions and provision of advice: applied mathematicians, statisticians, biologists, oceanographers, economists, social scientists, lawyers (even judges), civil service managers and bureaucrats, industry, ENGOs, journalists and politicians. Many of these can be acting as advocates for certain interest groups.

At the end of all this, the decision-maker is frequently left with a difficult task, namely to make sense of available scientific evidence as well as the information provided by other groups, to relate it to the interests of various stakeholders, and to shape a decision that is, ideally, both epistemologically and politically robust (Lentsch & Weingart, 2011).

Basic advice and decision structures in fisheries management

Generally, these structures have four levels, though those can differ in detail from one dispensation to another.

1. *A technical team* of scientists (usually mathematical specialists) who conduct the basic calculations for catch limits – these calculations are termed 'assessments' of the resource.
2. *A scientific committee* of scientists with a broader range of specialities (occasionally some other stakeholders too), responsible for developing scientific recommendations for management measures (including catch limits in particular).
3. Some form of '*intermediary group*' that may assume many forms, but essentially reflects the interface at which scientists, managers and stakeholders (such as industry and ENGOs) will exchange views on the scientific committee's recommendations; these recommendations are then forwarded (perhaps amended and likely embellished) to decision-makers.
4. *Decision-makers* who are seldom scientists, and are responsible for making final decisions on management measures.

Note that it is at the third of these levels at which the effective communication of science and the results it provides is of the most importance, if that science is going to have its appropriate impact on final decisions.

South African structures

Local structures differ somewhat from the general form set out above.

1. *A technical team* of scientists who may be drawn from government agencies, academics or freelancers on contract, and industry; such groups develop the assessments.

2. *Government department scientific working groups* (typically one for each fish species group, for example, for small surface shoaling fish including sardine and anchovy). The (voting) members of these groups are scientists, mainly drawn from the government department responsible, but observers representing stakeholders (such as industry and ENGOs) are generally permitted and participate quite fully in discussions. These groups develop scientific recommendations for management measures.
3. An *intermediary discussion process* within the responsible government department. This process is primarily internal amongst civil servants, though at times includes other stakeholders.
4. *Decision-makers*, usually the minister of the government department responsible, though the minister sometimes delegates this responsibility to the deputy director-general (DDG) responsible for fisheries. Final decisions on management measures are made at this level.

Effective communication (to non-scientists, i.e. laypersons) of science and the results it provides becomes important here at the second level. But it is at the third level where this communication process may often matter most, and unfortunately may frequently be poor because of under-representation of scientists. In other words, preparing the decision may become over-politicised at an early stage at the expense of giving due weight to scientific evidence.

How well are these structures working?

Over the first 15 years of this century, there was only one instance of a non-trivial change made at a higher level to a scientific working group recommendation for the catch limit for a major South African fishery. Probably this is a record second to none elsewhere in the world.

While communication of science in the scientific working groups has not been perfect, it has been adequate. Often the standard scientific conference style of presentation of results is used, though in a more interactive and conversational format with much more intensive discussion than customary at scientific symposia. Understanding of those results has been aided by most of the observers either having a scientific background or being highly qualified senior executives from major companies.

However, starting in 2016, a highly problematic situation developed regarding the West Coast rock lobster (known colloquially as *kreef*) fishery. By way of background, this resource is highly depleted, primarily as a result of the very heavy exploitation that took place in the 1950s and 1960s. The resource's abundance at present is estimated to be only some 2% of what it was about 100 years ago before large-scale harvesting commenced. The harvesting policy had been one of slow rebuilding of the resource (a more rapid rate of rebuilding would have necessitated sharply reduced catch limits and consequent socio-economic hardship). This lobster species, being relatively easily caught close to the coast, is also an important component of the government's small-scale fisheries policy, which aims to empower marginalised coastal communities by granting fishing rights to co-operatives within those communities.

The year 2016, however, brought a marked change in perceptions of this resource and fishery in general. There was strong evidence of a recent marked reduction in the abundance of the resource, particularly in the important Cape Peninsula region. This was coupled to evidence of substantial increases in poaching (illegal fishing) in this region. The scientific working group recommended a marked reduction in the allowed catch limit so as to prevent further reduction in abundance and restore sustainable utilisation. But the decision-maker – the DDG responsible for fisheries – decided to maintain the existing catch limit (see Rogers, 2018).

This sequence of events was repeated in 2017. As a result, in July 2018, WWF instituted litigation seeking that future West Coast rock lobster catch limits be set consistent with sustainability. Two months later, Justice Rogers issued a landmark and precedent-setting judgment (in that it applied also more widely to other renewable resources and to decisions relating to them as well) in the Cape High Court (Rogers, 2018). He found that the DDG's catch limit decision the previous year had violated the South African Constitution and the relevant national environmental and fisheries laws, and had been irrational. He also emphasised the need for such decisions to be based on the best available scientific evidence.

The government subsequently reduced the West Coast rock lobster catch limit for the 2018/19 season in line with a two-year step-down process, as recommended by the scientific working group. The minister and DDG initially gave notice of intent to seek leave to appeal the judgment (e.g. see Nkwanyana, 2018a). This gave rise to concerns that if the appeal were successful, some South African marine fisheries products might no longer be acceptable for import into certain countries with strict provisions that fish products come from sustainable fisheries. However, in December 2018, the government changed its mind and decided not to seek leave to appeal the judgment.

In terms of science communication, the chief concern to which this sequence of events gave rise was that evidence provided to the court by the DDG (see Rogers, 2018), together with press statements made by the minister's spokesperson (Nkwanyana, 2018a, 2018b), demonstrated an absence of fundamental understanding of the scientific concepts underlying sustainable management of renewable resources.

International structures

Fisheries that take place in international waters (outside national 200 nautical mile exclusive economic zones), or in the national waters of more than one country, are generally managed under Regional Fishery Management Organisations (RFMOs). The most important of these in terms of high-value fisheries are the five tuna RFMOs, including, for example, the International Commission for the Conservation of Atlantic Tunas (ICCAT). Their structures mostly follow the general form set out above quite closely.

1. *Scientific sub-committees* consisting of scientists who conduct assessments.
2. *A scientific committee* of scientists, generally (but not always) restricted to the nominees of member governments, who may also include fishery managers and/or other stakeholders. These develop scientific recommendations for management measures.
3. Some form of '*intermediary group*' comprising scientists, managers and stakeholders who may modify and/or extend the recommendations to be made to decision-makers (the Commission itself).
4. *The commission* consisting of one commissioner (usually a senior civil servant) from each member state who vote on the final decisions to be made.

RFMOs, which typically try to operate by consensus, often experience many difficulties in discussions and reaching such decisions. For example, most have many member countries (often in the dozens). There are usually very different levels of ability and experience amongst the scientists and amongst the managers from these different countries. This can make a consensus on other than 'no change' (status quo) to catch limits difficult to achieve, and this has at times led to poor performance by these organisations in managing their fisheries.

Commissioners (and also government fishery managers) are typically more skilled at negotiating (involving, for example, sharing an overall catch limit), but less comfortable when participating in scientific discussions.

A form of 'competition' can arise with other international organisations which have partly overlapping responsibilities. One such example is the Convention for International Trade in Endangered Species (CITES), particularly as national delegations to RFMOs tend to be dominated by representatives from fishery departments, while those to CITES tend to come primarily from environment departments, and these two groups often have conflicting views on utilisation versus species protection trade-offs.

An attempt to improve this situation

Recently, a broad initiative has developed to attempt to improve this situation by adopting what is called a 'management procedure' approach to recommending catch limits in these organisations (Butterworth, 2007; Punt et al., 2016). The approach was first developed in the scientific committee of the International Whaling Commission in the late 1980s. The primary motivation behind this approach was the simulation testing of proposed formulae to set catch limits so as to ensure that they were appropriate in the face of uncertainty about resource abundance and productivity, i.e. even if best perceptions about the resource were wrong, the formulae would self-correct adequately to ensure the resource was safeguarded, hence taking due account of the precautionary principle (Anon., 1992).

The approach has since been adopted in, for example, some national fisheries in South Africa, Canada and New Zealand, and internationally by the North Atlantic Fisheries Organisation for Greenland halibut and by the Commission for the Conservation of Southern Bluefin Tuna for that species (see Punt et al., 2016; Nakatsuka, 2017).

A key aspect of the approach is that the data inputs and formulae to be used in calculating catch limit recommendations for the next few years are *pre-agreed*; in other words, the rules are agreed before the fisheries management game is played out. In this way, it is hoped that consensus can be built around the catch limit recommendation arising from this process, rather than reverting to the status quo, so that catch limits will change in scientific accordance with the changing status of a resource.

A positive development in this process of achieving wider usage of the management procedure approach was the agreement by all five tuna RFMOs that they would all move in the direction of setting catch limits on this basis in the future (Anon., 2011). However, progress is not proving as rapid as initially hoped, with one somewhat surprising reason being advanced that it is scientists themselves who constitute the major stumbling blocks to the process. Complaints include the fact that there is a lack of commonality in explanations given and material (such as tables and plots of results) provided by scientists in different RFMOs, which confuses stakeholders (particularly government managers) who often attend meetings of more than one of these RFMOs. There are complaints that many scientists are themselves not well versed in the concepts that underpin this new management procedure approach.

Essentially then, it is scientists' poor ability to communicate with stakeholders from outside their scientific discipline that is argued to be the root of the problem. This realisation led, in turn, to an initiative launched by the PEW Organisation in 2017 to improve the situation, as discussed in more detail below.

Addressing problems in communication between scientists and stakeholders in fisheries

A scientist's basic training runs along the lines that any conclusion (and related recommendation) put forward must be prefaced by statements of the assumptions made, and a full explanation of the underlying analysis to provide defensible justification for the advice provided. Often scientists' expectations of stakeholders and decision-makers is that they will interact on the basis of this scientific paradigm, and will have the time and interest to participate fully in the associated scientific discussions.

Generally, however, stakeholders and decision-makers simply do not follow the scientific paradigm. Many scientists need to better realise that presentations of their arguments must be styled to target their audiences, which in fisheries will often consist primarily of laypersons. A senior official in the Australian southern bluefin tuna industry, also with considerable experience in negotiations with politicians in that country, advises that Australian government cabinet ministers generally want to hear no more than the recommendation itself with possibly a soundbite on the underlying rationale. This preference on the part of politicians has been aptly expressed by the famous quote attributed (amongst others) to President Truman: 'give me a one-handed economist'.[2] Personal experience of international courts has indicated that (unsurprisingly) the primary ability required by counsels in presenting their cases in these fora is to style their presentations in a manner that will maximise understanding by and impact on the judges.

Essentially therefore, when fisheries scientists present to primarily lay audiences, they need to be able to reverse the standard scientific approach which they were taught. Hence present 'top-down' rather than 'bottom-up', i.e. start with the conclusion, and follow that with the essence only of the justification for that conclusion, expressed in laypersons' language.

This was the main message to emerge from the PEW organisation's initiative mentioned above. In January 2017, they organised a workshop including primarily scientists and stakeholders from RFMOs which had successfully implemented the management procedure approach to try to distil what had been the main reasons behind those successes. The outcome is reported in Miller et al. (2018). The workshop's primary recommendation was for a greater focus on meetings of 'intermediary groups' to allow for improved scientist–stakeholder interactions, together with the development of improved visual communication tools for the presentation of what are often quite complex results.

Conclusion

To summarise on the key science communication needs for fisheries, in South Africa these would seem to involve improved interactions between scientists and decision-makers. The recent judgment by Justice Rogers in the Cape High Court (see above) should assist, given the emphasis it placed on the important role for science in management decisions for fisheries.

More generally, there is a need for scientists to better style and focus their presentations and interactions for primarily layperson

2 https://quoteinvestigator.com/2019/04/10/one-handed/

audiences. An increase in the interaction between scientists on the one side, and stakeholders and decision-makers on the other, in 'intermediary groups' within the fisheries management decision structures, would also be advantageous.

It is evident that the general guidelines for science communication to the general public are not the most appropriate in contexts where scientific evidence is needed to inform decisions which have direct political and/or economic consequences, or may adversely affect the natural environment. While the effects of science communication to the general public are rarely ever evaluated, the communication of scientific evidence or the lack thereof to stakeholders or decision-makers are evaluated by implication if decisions are based on this.

Acknowledgements

I thank Peter Weingart for helpful suggestions on an earlier version of this chapter.

References

Anonymous (1992). The Rio Declaration: Principle 15: the precautionary approach. http://www.gdrc.org/u-gov/precaution-7.html.

Anonymous (2011). Kobe III recommendations, http://tuna-org.org/Documents/TRFMO3/K3-REC ENG.pdf.

Butterworth, D. S. (2007). Why a management procedure approach? Some positives and negatives. *ICES Journal of Marine Science,* 64: 613–617.

Lentsch, J. & Weingart, P. (2011). Quality control in the advisory process: towards an institutional design for robust science advice. In J. Lentsch & P. Weingart (eds), *The Politics of Scientific Advice: Institutional design for quality assurance* (pp. 353–374). Cambridge: Cambridge University Press.

Miller, S. K., Anganuzzi, A., Butterworth, D. S., Davies, C. R., Donovan, G. P., Nickson, A., et al. (2018). Improving communication: The key to more effective MSE processes. *Canadian Journal of Fisheries and Aquatic Sciences*, 76, 643–656. doi: 10.1139/cjfas-2018-0134.

Nakatsuka, S. (2017). Management strategy evaluation in regional fisheries management organizations: how to promote robust fisheries management in international settings. *Fisheries Research,* 187, 127–138.
Nkwanyana, K. (2018a). Ministry explains the West Coast rock lobster total allowable catch and effort reduction. Media release, 19 November. South African Department of Agriculture, Forestry and Fisheries.
Nkwanyana, K. (2018b). Statement by the Minister for Agriculture, Forestry and Fisheries, 2 August. South African Department of Agriculture, Forestry and Fisheries.
Punt, A. E. (2006). The FAO precautionary approach after almost 10 years: Have we progressed towards implementing simulation-tested feedback-control management system for fisheries management? *Natural Resource Modeling,* 19(4), 441–464.
Punt, A., Butterworth, D. S., De Moor, C. L., De Oliveira, J. & Haddon, M. (2016). Management strategy evaluation: Best practices. *Fish and Fisheries,* 17, 303–334: doi: 10.1111/faf.12104.
Rogers, O. L. (2018). Judgment in the High Court of South Africa (Western Cape Division), Case no. 11478/18.

6 Science and social media: Opportunities, benefits and risks

Shirona Patel

Making science accessible: A new mandate

Scientists in open democratic societies are under increasing strain from the state, funders and other societal actors to make scientific knowledge accessible, to decommodify knowledge and to conduct research that impacts on society and contributes to the global knowledge economy. Scientists are also required to demonstrate accountability, especially when funded by the public coffers.

This paradigm shift accelerates the demand for new knowledge and scientific content to be made visible in the public sphere (Badenschier & Wormer, 2012; Pavlov et al., 2018), with a growing demand for science communication and engagement efforts from the research community (Pavlov et al., 2018).

Indeed, funding policies in many countries around the world, including Australia, China and South Africa (Joubert, 2019), make science engagement mandatory for researchers and institutions funded by the state. For example, the National Aeronautics and Space Administration (NASA) is an independent agency of the US government, governed by the National Aeronautics and Space Act[1] which stipulates that NASA is obligated to 'provide for

---

1 National Aeronautics and Space Act (2010), https://www.nasa.gov/offices/ogc/about/space_act1.html and and https://www.nasa.gov/audience/formedia/features/communication_policy.html

the widest practicable and appropriate dissemination of information concerning its activities'. Similarly, the multi-node Centres of Excellence and other researchers, centres and chairs funded by the South African National Research Foundation (2018) are often obliged to dedicate a portion of their grants to science engagement. These are global and local examples of how science is funded by the public and how making research accessible to the public is part of the mandate of researchers (Pavlov et al., 2018).

Similarly, more and more individual philanthropists, corporates, private sector funders and trusts and foundations are insisting that scientists make evident the impact of their studies and that they engage with a range of publics to make their work visible. A case in point is the Wellcome Trust which under its Public Engagement Fund allocated specific funding for the use of creative approaches for this purpose (Wellcome Trust, n.d.).

In the *2018 State of the Newsroom Report*, Kruger (2018: 2) writes that there must be a balance between academic rigour in research and making the research accessible to the public: 'It is no longer feasible for a university-based journalism programme to lose itself in purely academic research.'

Researchers at universities and knowledge-based institutions are encouraged to explain the impact of their research on society, whether it be through discovery research that changes disciplinary thinking, translational research that influences policy and practice, or innovative research that can be taken to the market to generate economic activity. For example, a group of interdisciplinary earth scientists who work for the Ocean and Sea Ice section of the Norwegian Polar Institute (OSINPI) believe that through actively communicating the results of their studies, and sharing new knowledge based on evidence in the public domain, they contribute to addressing the deficit in fact-based knowledge around climate change. They use social media to continuously share information about their research and matters related to climate change, thereby trying to effect real transformation of thought in society based on scientific proof. As a small group of young scientists and researchers with limited resources, they use different social media strategies to facilitate multi-way communication with a variety of publics, including fellow scientists and collaborators, policy-makers, funders, the media and the general public. This example provides evidence that researchers and scientists can successfully use social media channels to make knowledge visible globally, through limited resources, without relying solely on professional science communicators.

Traditional media in decline

Scientists often use the mainstream media as a conduit to reach multiple publics (Joubert & Guenther, 2017).

The media has an essential role to play in open progressive democracies to develop an informed public (Dahlgren, 2009; Gumede, 2014), amongst other priorities. However, the traditional print media and some broadcast media in South Africa are under severe economic strain due to the advent of new digital technologies and platforms, changing patterns of media consumption, declining print circulation, the closure of newspapers, and the introduction of new business models (Breitenbach, 2019; Kruger, 2017). Finlay (2018: 3) describes the 'dissolution of "the newsroom" as we know it', evidenced by the closing down of many print titles and widespread retrenchments in both the print and broadcast media in South Africa, including the proposed retrenchment of over 900 staff at the South African Broadcasting Corporation (Finlay, 2018).

The 'decimation' of newsrooms (Daniels, 2018a), the 'integration' of editorial and commercial activities (Cornia et al., 2018) and the decline in the number of specialist journalists assigned to specific beats, like science, health and education, has long been lamented (Daniels, 2018a; Thloloe, 2005), with the general quantity and quality of science reporting found to be inconsistent, unstructured and relegated at the expense of more newsworthy genres like politics and economics (Claassen, 2011; Van Rooyen, 2002). The number of specialist science journalists in the traditional media is diminishing with less than ten permanent science journalists in South Africa in 2018 (South African Science Journalists' Association, 2018). Experienced journalists are being laid off and the degeneration of beat journalism is a global issue (Daniels, 2018a). Daniels (2018b: 4) adds that 'many retrenched journalists go into the gig economy, including doing public relations, scratching out an odd-jobs living'.

Due to resource constraints, there is a real risk that new, important scientific research may be ignored and that society may remain in the dark regarding innovative scientific developments (Badenschier & Wormer, 2012). Limited resources often result in the lack of capacity to proactively pursue stories; to report fairly, accurately and credibly; to fact-check; to explore multiple angles of an issue; and to properly investigate important, relevant viewpoints pertaining to a specific matter.

In a 'fake news' and 'post-truth' environment (Finlay, 2017), where 'science is vulnerable to abuse and distortion, especially for political purposes' (Kizer, 2018: 1) including by other scientists (Peters, 2013), in a setting where false information, science quackery and 'information disorder' is on the increase (Bourguignon, 2018), it is imperative to understand how science is sourced, assessed, selected and published, and for whose benefit. Given the constraints facing newsrooms, including the increasing power of commercial actors, advertisers, audiences, media owners, politicians and sources (Stromback & Karlsson, 2011), and the fact that journalists and editors are under significant pressure to publish new content to feed the 24-hour news cycle, there is a need for a steady flow of reliable information to newsrooms (Schudson, 2003).

As fake news proliferates, trust in the traditional media is declining, and people are becoming less believing of established sources, whilst appreciating the influence of their peers (Broniatowski et al., 2018; Hetherington in Hart & Shaw, 2001; Jones, 2004).

Understanding the tensions between scientists and the traditional media

The media creates distance between scientists and publics, does not contribute enough to the public understanding of science and does not elaborate on the impact of science on the daily lives of people, according to Nelkin (1995: 14–15) who writes that science appears in the press as 'an arcane and incomprehensible subject'. Whilst this view was expressed over two decades ago, there is still dissonance between how scientists communicate within the scientific community versus how scientists engage around scientific matters in the public domain (Peters, 2013).

Wilcox (2003) claims that science journalism norms do not sit comfortably with those of the science being covered. She states that science journalists need conflict, drama or exclusives to make science appealing to news editors, whilst scientists de-emphasise single studies and rather promote the full body of science in context. This tension is also identified by Lynch and Condit (2006) who expand on the tension between journalists who need to make stories interesting and 'sellable', on the one hand, and scientists who want stories to be neutrally reported, balanced and accurate. The lack of control over the journalistic process is identified by Peters (2013) as a major hurdle in the relationship between scientists and journalists, with some researchers opting to work with alternative models like *The Conversation* where they have final sign-off on articles before publication.

There are two major factors confronting science journalism according to Cornelia Dean, the former news editor of the *New York Times* (Dean, 2002). She claims that science journalism's reach has to be very broad, yet science is becoming increasingly specialised, so journalists cannot keep up in an age where scientific research is becoming more commercialised. Hotz (2002, in Badenschier & Wormer, 2012) believes that the relationships between science journalists and scientists 'is becoming increasingly fraught', a tenet supported by Claassen (2011).

Whilst the pursuit of the truth is a value that forms the basis of both the journalistic and scientific fields, and whilst both journalists and scientists seem to embrace the shift to an open, transparent society, given the waning trust in the traditional media (Edelman Trust, 2018) and the difficult relationships between scientists and journalists (Claassen, 2011), scientists are gradually employing direct, digital communication strategies to make science accessible to multiple publics, thereby discounting the reliance on the traditional media (Fuchs, 2014; De Lanorelle, 2017; Daniels, 2018).

Scientists are becoming increasingly skilled in media management (Franklin, 2004) and are progressively relying on professional science communication practitioners (science communicators) to share and amplify science in order to make their research relevant and visible in the public sphere (Kiernan, 2006; Stromback et al. 2012). A study focusing on South Africa's most vocal scientists (Joubert & Guenther, 2017) reveals how scientists who are also good science communicators emerge in the news and are more popular. The following science themes are covered the most in the South African print media: environment and ecology; health sciences; science and technology; zoology; astronomy; energy; anthropology and archaeology; engineering sciences; the palaeosciences; food and nutrition sciences; and physics (Patel, 2019).

The general quantity and quality of science reporting is inconsistent, unstructured and relegated to the middle pages of newspapers at the expense of more newsworthy genres like politics and economics (Claassen, 2011; Van Rooyen, 2002). Resource, time and space constraints, the declining number of specialised science journalists and the need for science journalists to write across titles and platforms in real time to feed the ongoing digital news cycle are some of the factors that impact the publication of science in the South African media (Patel, 2019).

However, despite the reduction of the number of dedicated science desks and specialised journalists, a three-month study of South African print newspapers (Patel, 2019) reveals that science made it to the front pages of two newspapers a total of eight times during the period, with four newspapers publishing editorial columns on science-related issues.

Scientists are thus faced with the quandary as to whether to use the traditional media as a conduit to reach wider publics, whether to develop their own virtual communities or whether to use a combination of the traditional and social media.

Why should scientists communicate?

There are several reasons why scientists communicate and why they should communicate. People communicate to share information; to persuade others to do something or to change their perceptions or behaviour; to express their opinions on a particular matter; to commit to doing something; and to transform society (Searle, 1979). According to Gascoigne and Metcalfe (2012), scientists communicate to create awareness, to add value to the public discourse, to start a conversation, to share insights from their research that may be beneficial to broader communities and to create impact in society. Scientists also communicate to engage with publics, to obtain feedback on ongoing research and to serve as a catalyst for social change.

There is a need for scientists to build relationships and foster collaboration within and across research areas, as universities and research institutions encourage inter-, trans- and multi-disciplinary studies across disciplines, faculties, universities, institutions and sectors. Collaboration and inter-disciplinary research are recurring themes in the South African White Paper on Science Technology and Innovation (DST, 2019), which emphasises that talent and resources available in universities and research entities, coupled with industry support, should be harnessed to ensure that South Africa is prepared to actively participate in the Fourth Industrial Revolution (DST, 2019).

From a public relations perspective, the benefits of communicating science include enhancing the reputation of an individual researcher, or a team of researchers, attracting collaborators, students and programmes, and securing funding for research projects.

An important role of science communication is to influence policy in a country or indeed across nations. The influence of the Treatment Action Campaign and other activist organisations that used both traditional and new media forms, combined with strong advocacy and lobbying tactics, to pressurise the state into providing antiretrovirals to people living with HIV/Aids in South Africa, is well-documented in *Reporting the South African HIV Epidemic* and other studies (Muchendo, 2005; Palitza et al., 2010). This is a pertinent example of how science communication can help to effect real change in society, and in this case, result in the saving of millions of lives.

Another successful example of where prolonged science communication and advocacy has influenced policy is evident in the implementation of a new 'sugar tax' on sugar sweetened beverages that was legislated in South Africa in 2018, following the implementation of such a tax in Mexico, Chile, Denmark, France, Hungary and several other countries (Stacey et al., 2017).

In the face of tremendous pressure from the beverage industry and amidst threats of job losses and intimidation on a number of fronts, the ongoing science engagement efforts of members of a research unit known as PRICELESS SA (Priority Cost Effective Lessons for System Strengthening South Africa)[2] based in the School of Public Health at the University of the Witwatersrand, enabled the team to empower both citizens and policy-makers with the relevant information based on research, using the media and other advocacy initiatives, to make decisions about health investments in South Africa. PRICELESS SA also provides scientific information that seeks to improve the way in which resources in the country's health and related budgets are allocated and priorities are set to improve public health.

In the example described above, PRICELESS SA faced numerous challenges from the local and international beverage industry and some unions, had to contend with massive misinformation and disinformation placed in the public realm, and had

---

2 www.priceless.co.za

to ward off multiple threats in public and private, in their quest to impact on policy in South Africa. However, there are other instances where science communication has been less effective in changing policy, or where individuals, scientists, lobby or advocacy groups communicate to further their own agendas.

Similarly, the implementation of policies related to vaccination, tobacco, rhino poaching and energy are often made controversial in the public space, not always through a deliberation based on science, but often through the way issues are themed in the media. For example, the proposed Control of Tobacco Products and Electronic Delivery Systems Bill of South Africa (2018), which seeks to regulate the tobacco industry (including e-cigarettes and vaping products) and to remove branding on all tobacco products at point of sale, resulted in a major controversy in the media between tobacco manufacturers, the producers of e-cigarettes and vaping products, trade unions and pro-choice lobby groups, on the one hand, and the national department of health, and the National Council Against Smoking on the other.

Changing news values in a digital world

There are major shifts reshaping the science journalism landscape with the impact of new media technologies in a changing media environment recasting science journalism's familiar norms and values in unanticipated ways (Allan, 2009).

The digital disruption that we experience today impacts the news values and indices that influence what news is published, how it is assessed, selected and framed, who influences the publication of science news, and which platforms are selected for publication. According to O'Neill and Harcup (2009), it is essential to study news values because it 'goes to the heart of what is included, what is excluded, and why' some news is given precedence over others.

Badenschier and Wormer (2012) describe news values as factors that make news valuable and add that the value of news increases based on the number of news factors present in the article as well as the intensity of these factors. They attempt to develop a science news index in 2012, with specific criteria being developed to determine what makes science newsworthy, an index which is still in development. They found that with regard to the selection of science in particular, that 'graphical material' was an important factor in selecting science news for publication and that editors had to not only select the news but also to consider the platform through which the news would be published, an additional factor that influences what becomes news in a digital era.

Whilst several news value indices have been developed over the last five decades, Harcup and O'Neill (2001) revisited Galtung and Ruge's (1965) list of ten news values and developed their own list of ten factors that make content newsworthy. They claim that particularly good and bad news make the news as do the following: news that is significant in magnitude and relevant to audiences; stories with an element of surprise; entertaining stories that focus on the powerful, the elite or on celebrities; follow-up articles; and those that fit the newspaper's agenda. Their most recent list of contemporary news values (Harcup & O'Neill, 2016) is adapted to accommodate digital and social media with the following five news values added: exclusivity; conflict; the use of audio-visual materials; shareability; and drama.

People share content depending on the news values contained in the post. Social media posts that are relevant, unexpected, and that include some form of controversy or negative consequence, and that may potentially impact on many people, are more likely to be shared (Rudat & Budar, 2015).

Social media lends itself to participatory science and to empowering citizens

Despite science and society moving closer together (Weingart, 2001, in Hargittai et al., 2018), there is limited research on how researchers and scientists use social media to communicate science (Hargittai et al., 2018), how users engage with science, research and new knowledge through scientific content in the digital sphere; and how scientists interact with their peers, the public and other users via online and social media.

Science communications developed as a professional field after the Second World War with science communication models evolving over the last seven decades. The initial 'public understanding of science' model assumes that the public's knowledge of science is deficient because the public does not understand science, and scientists thus have to bridge this knowledge deficit by informing and educating the public through the use of unidirectional mass communication tools, in which the public is a passive receptor of information (Peters, 1996, in Hargittai et al., 2018).

A second model focuses on 'public consultation', which sees scientists engaging with the public to obtain their views on a particular issue like the efficacy of vaccines or their views on climate change. In this model, the public provides feedback on a science theme, topic or issue that is actively placed into the public domain by scientists. This could take the form of a seminar, public lecture, workshop or conference. New digital technologies like online surveys can be used to obtain the views of members of the public, for example, on new science that has been shared in the public domain. Social media polls are one way of gauging the public's response to scientific matters but are not representative samples that can be used for scientific purposes.

Newer science communication models speak to 'participatory science' which involve multi-way communication with various users, including members of the public, who, despite being non-experts, help to set the agenda, make decisions, and influence policy and knowledge production processes (Bucchi & Neresini, 2008).

It is this latter definition that aligns most with the participatory digital technologies of today (Hargittai et al., 2018). Social media lends itself to participatory science (Brossard, 2013) because there are low barriers to engagement provided that one has access to data and the internet which are still impediments in some developing countries like South Africa (Hootsuite, 2018); the playing fields are levelled for producers and users of content, and all parties have the ability to create, share and exchange information, ideas and content on similar platforms in real time.

Access to the internet and data remains a major global hurdle more than half of the world's population still not online (WEF, n.d.). The WEF (n.d.) identifies four barriers to internet inclusion: infrastructure; affordability; skills, awareness and cultural acceptance; and relevant content'. A two-year enquiry into the cost of data was undertaken by the Competition Commission in South Africa and the preliminary results reveal that the cost of data is much higher in South Africa than in peer countries (ICASA, 2019).

On social media, users have the power and the ability to decide how they would like to interact with the content and fellow users, how to filter and to manage what information they would like to receive and whether they would like to share or amplify content. Users can also control who they connect and interact with and how they engage with the content of others, for example through sharing content, liking content, or commenting on the content. According to Hargiatti et al. (2018), social media thus enables engagement though content and human interactivity, all of which can increase the number of participants engaged with science.

The terms of the debate have changed with an exponential rise in the adaptability of platforms that enable citizens and empower stakeholders to help create and reshape the news in the digital sphere (Hamshaw et al., 2017), without the reliance on the traditional media to serve as a conduit to the general public.

For example, #EarthHour[3] is widely known as an annual project of the World Wildlife Fund that aims to get people from across the world to shut down all electrical appliances for an hour in order to raise environmental awareness globally. In 2019, this campaign reached over 188 countries around the world, in part due to the impact of social media.

Social media and the internet have also transformed the conceptual framework in which people interpret, perceive and respond to risk (Chung, 2011). Social media platforms provide quick access to information in real time, serve as a sounding board

3 www.earthhour.org

and a content hub for a range of questions, proffer the opportunity to create and develop virtual communities, and enable users to connect with those of similar views (Flanagin et al., 2014).

The availability of smartphone applications and access to mobile data has changed how people search for, access and consume information in real time. At the same time, in a digital world where fake news, bots, trolls and malware have the ability to harm, to spread unsolicited content, to promote discord, and to create false equivalency (Broniatowski et al., 2018), it is essential to develop an informed digitally literate public who are savvy enough to see through misinformation and disinformation online, who can read the context within which information is shared and question the sources of the content. It is fundamental to develop digitally literate individuals who are agile enough to comprehend how issues are framed online and to understand whether fellow users are real or not. This is not an easy feat, as some social scientists believe that 'scientific knowledge both embeds and is embedded in social identities, institutions, representations and discourses' (Jasanoff, 2004: 3). If this is indeed the case, there is a risk that scientists try to order the world through how they understand it, thereby trying to regulate and systematise it according to their own views or findings (Jasanoff, 2004).

Broniatowski et al. (2018) describe the use of bots, trolls and malware to sway public perception in the vaccine debate online. See details in the section on social media risks below.

Science and social media

In the context of science engagement, social media can be described as digital networked communication channels that allow for information to be accessed and shared, and interactions to be facilitated amongst and between researchers, scientists and fellow knowledge workers, as well as with multiple other publics in the digital sphere (Collins et al., 2016). Social media can also serve as a 'complementary information network for individuals who consider being well informed as highly important' (Kuttschreuter et al., 2014: 10).

How are scientists using social media?

In a study involving 587 scientists worldwide, Collins et al. (2016) concluded that scientists across disciplines, faculties and institutions are using social media platforms to exchange scientific knowledge, although very few scientists are engaging in social media. Scientists are also using these channels as open, multi-way channels, to communicate particular aspects of research and science as a means of outreach, to increase science engagement and to encourage science literacy in society.

Collins et al. (2016) found that Twitter, Facebook and LinkedIn were used by the majority of scientists surveyed to share research and new knowledge, as well as Instagram. Science blogs were viewed by the majority of the respondents in this study (84%) to be an important online platform for science engagement.

Respondents used Facebook to share experiences in the laboratory or field, to find inspiration for outreach and science communication, to connect with fellow researchers and to correct fake science news in the public domain. Whilst this study found that Facebook could play a role in bridging the knowledge deficit and encouraging consultation on a particular topic, 'not many respondents found Facebook to be a suitable platform for discussion or to develop scientific literacy' (Collins et al., 2016: 5). They also did not find Facebook useful for communicating with the general public or with fellow researchers.

However, the study found that scientists spent between 15 and 60 minutes on Twitter every day on 'scientific tweeting', described as a tweet based on a science subject, created or shared by a scientist, that usually included a science-related hashtag. These scientists were found to tweet about research within their own field, particularly when they were participating in a conference or event, where they generally used the hashtags created by conference organisers. Scientists in this study found Twitter to be a useful medium for engaging with fellow scientists, the public, other audiences and the media.

The majority of scientists in the Collins study found science blogs to be informative, with 89% of the scientists surveyed agreeing that blogs were valuable in explaining science to the public. About half of the respondents claimed to have written their own blogs (Collins et al., 2016).

This study differed from the results of a US study into how young adults use social media for science communication conducted by Hargittai et al. (2018), which found that most young adults in the US used the Facebook platform, followed by Twitter, with almost 40% of the respondents using both channels. They found that 44% of young adults shared science and research content via Facebook compared to 10% of Twitter users. The study also revealed that Twitter was a popular medium through which to share science and research content, particularly during and after events and conferences.

A slightly different approach was adopted by Pavlov et al. (2018), a group of researchers from the Ocean and Sea Ice section of the Norwegian Polar Institute (OSINPI), who provided an account of how they have successfully used social media over a three-year period to reach young audiences through an essay in the *Bulletin of the American Meteorological Society*.

Comprised of about 20 members, the OSINPI group engages with fellow scientists and the general public through Instagram, Twitter and Facebook. Instagram was deliberately selected as the first medium of choice by the team as it is a visually appealing medium that attracts younger audiences and connects with younger people emotionally. Instagram lent itself to the project, as the OSINPI group had a range of good quality visuals to share, with fieldwork and educational posts proving to be popular content. Selected posts were amplified through collaboration with similar entities and influencers like *National Geographic*, who were tagged in some posts and who shared some of the content via their respective social media channels.

The second medium of choice for the OSINPI group was Twitter, selected because of its appeal to older, engaged audiences that included members of the media, fellow researchers and scientists, influential politicians, policy-makers and consultants. The

Twitter posts contained new information and breakthroughs and the platform was particularly used for live tweeting from events and announcements. The use of hashtags and keywords to better engage with the public was also a successful strategy employed by the group.

Facebook was also included in the OSINPI social media strategy due to its wide reach and ability to connect personally with colleagues, friends, and like-minded groups. Posts related to the achievements of scientists, or news related to researchers, including profiles, were shared the most on Facebook.

It is important to distinguish content by platform (Yeo, 2016). Whilst Twitter may be good for expert debates about science and research, Facebook may be a better medium to bring science closer to selected communities, whilst Instagram may encourage visual engagement. In all instances, it is important for the content to be captivating and tailored to the different audiences using these platforms.

Pathologists are also using social media for collaboration and networking purposes and for sharing information with the general public. This is according to a study by Gardner and McKee (2019), which stipulates that there are more than 4 700 pathologists and pathology-related accounts on Twitter. However, pathologists are also using Facebook to share educational content like useful case studies, resources and articles, with in-depth discussions on particular cases taking place in Facebook groups. Similar usage patterns can be observed in the Early Southern Sapiens Facebook study group,[4] an online community led by Professor Christopher Henshilwood and comprising of hundreds of scientists, researchers, communicators and interested parties from around the world, with the primary objectives of trying to establish when, why and how humans first became behaviourally modern and what it means to be human.

Similar to the OSINPI group, selected images are shared by pathologists on Instagram, and Twitter is used for sharing

4 https://www.facebook.com/groups/SouthernSapiens/

content from meetings and events. However, Gardner and McKee (2019) emphasise the benefits of using YouTube for teaching and educational purposes. They describe the use of this social media platform as transformative as it enables communicators to become more efficient as educators, provides the ability to share content across the world, even in 'medically underserved areas', and when incorporated into official curricula, allows the sharing of video content with students and colleagues, which frees up time to conduct further research.

The use of smartphones and 'digital photo-microscopy' allows for high resolution images to be taken with a smartphone and to be shared digitally in real time, and in so doing to traverse the barriers of time and distance when working on pathology cases that require immediate review (Gardner & McKee, 2019).

Whilst no comprehensive study exists in South Africa as to how researchers use popular social media, a study by Onyancha (2015) determined that scientists at research-intensive universities in South Africa who used social media platforms like ResearchGate were more likely to obtain coverage, to register a higher impact score and have their universities feature in the global university rankings.

In terms of popular social media, there are project-specific examples that offer some insight into how scientists use popular social media to communicate science. For example, a study by Mudde (2019) explores how South Africa's two most visible scientists, Professors Lee Berger and Tim Noakes interact on Twitter. The research establishes that both scientists try to be accessible and transparent and use Twitter to inform, educate and sometimes entertain their followers on matters related to their respective disciplines. In another instance, research into how the Square Kilometre Array (SKA) Telescope was represented on Twitter over a period of a year found that whilst most tweets were from large media organisations and leading science journalists, there were substantial opportunities for high-profile individual users to shape the discourse around the SKA (Gastrow, 2015).

Creating compelling social media content

Whilst scientific data related to social media metrics is freely available, the psychology behind why some users engage with some content and not others is scant (Hwong et al., 2017) and there is a need to research deeper forms of interaction, including why people click through to some articles and not others, or spend more time reading some blogs at the expense of others. Further, whilst social media enables users to create their own content, and interact with science content by retweeting, sharing, commenting and liking content, there is a need for research to examine how users engage with content and with fellow users online (Hargittai et al., 2018; Hwong et al., 2017). This extends to how users really engage with content, for example, by clicking through a link to find out more about the subject and also whether this engagement translates to influence or behavioural change over time.

On social media it is important to be authentic, different, respected and influential. The development of a unique persona and voice, coupled with humour and delivering content that people really want, are key considerations for developing any form of social media (Hootsuite, 2018). It is important to 'focus on creating mutual value instead of just trying to sell more stuff, make it easy for people to engage online and to use digital tools to keep the conversation going' (Hootsuite, 2018: 9).

Pavlov et al. (2018) advise that good quality audio-visual material, coupled with clear, concise, high quality, clever text, tailored per platform, make for good content to develop audiences. Content is key, should be planned in advance, and can include posts related to science education, laboratory or fieldwork, publications, team achievements, relevant events, breaking news, and historical posts like 'Throwback Thursdays' or 'Flashback Mondays'.

The ability of scientists to freely express their views, with little or no institutional limitations is also described as a key factor for a successful campaign by the OSINPI group. They also advise on working in teams, sharing experiences, and collaborating with communications units in research institutions and universities to develop professional communication strategies. They explore creating complementary social media accounts to amplify campaigns and to boost them where appropriate through page advertisements, promotions and other forms of paid-for content.

In 2017, Hwong et al. (2017) conducted psycho-linguistic research into why some space science posts are more appealing than others. They examined NASA's social media accounts (31 million Twitter followers and 21 million Facebook likes in 2019) and ascertained at the outset that space science is visually appealing and that the images and audio-visual material from space automatically lends itself to social media. However, they also found that aside from good, high-quality images, the visual description of the images, and content that evokes anger, authenticity and anxiety, makes for more engaging content on Facebook. In comparison, the top features for compelling and engaging space science in Twitter content are found to be visual elements like photographs, gifs and videos, and posts that include URLs and hashtags. Remarkably, Twitter posts that hint at some sort of 'certainty' are found to be more engaging by users.

The development of future social content will be rich content that includes social television for mobile devices, live broadcasting on social media, but it is all dependent on whether users have access to the internet and sufficient data. Access to the internet and affordable data was recognised and deemed to be a priority for South Africa to advance its developmental priorities according to the Minister of Communications and Digital Technologies, Ms Stella Ndabeni-Abrahams.[5]

An example of a social media campaign that delivered a high engagement rate[6] was a university-based campaign developed by communications professionals and scientists and doctors from the Donald Gordon Medical Centre in Johannesburg to announce

5 Minister of Communications and Digital Technologies, Ms Stella Ndabeni-Abrahams was speaking at the 2019 Digital Economy Summit in Johannesburg on 5 July 2019.

6 This refers to engagement with the content in the form of reactions, comments and feedback.

the results of the first intentional liver transplant from an HIV-positive mother to her child in October 2018.[7] The overall objective was to inform the public about the option of transplanting an HIV-positive liver into HIV-negative individuals in order to save lives and to encourage more people to donate their organs in South Africa. In total, 22 social media posts were developed for Facebook and Twitter, and together with a YouTube video, reached over 200 000 social media users. Of these, about 15 000 users engaged with the content within three weeks (measured through clicks, shares, likes and comments), with 58% of the users being female and 42% male. The engagement rate for the entire campaign was calculated at a rate of 6%, which is an extremely high rate for a social media campaign, as engagement rates for most campaigns average between 1% and 2% (Khumalo & Minors, n.d.).

The benefits of using social media

Professional science communicators, scientists and researchers, advance a multitude of benefits for using social media to communicate science.

Pavlov et al. (2018) view social media as an opportunity to bridge the gap between science and society, to engage the next generation of scientists, to reach out to the public and to empower policy-makers so that they can make informed decisions that will help to shape a better future for all.

Social media empowers citizens who are able to actively produce their own content, to engage with content that they want to receive, to curate content and to limit or expand on content. Users have the ability to decide on what content they want to receive and from whom, through which platform, and to become their own active gatekeepers as they select what content they want to share with their respective communities.

---

7 http://www.wits.ac.za/news/latest-news/research-news/2018/2018-10/worlds-first-intentional-hiv-liver-transplant.html

Social media also enables information to be shared in real time and facilitates rapid engagement on science and research between scientists and the public (Bik & Goldstein, 2013).

According to the respondents in the Collins et al. (2016) study, the ability to reach a wide, engaged and diverse audience, the ease of communicating in a short message format, the ability to project a view or share a message in real time and the accessibility of Twitter as a medium are some of the reasons why scientists feel that Twitter is an excellent medium through which to communicate science. The respondents in this study also claim to benefit from networking and communicating with other scientists through Twitter, as it provides access to issue publics that proffer multiple views from across the globe, which can be easily shared with fellow scientists, the public, science journalists and other relevant social actors. Twitter is also inexpensive and can be used in resource constrained environments (Pavlov et al., 2018).

However, the cost of interacting on social media, for example, through exchanging private information for the use of a platform or social media application, is viewed by many critics as a major risk borne by users who are often oblivious of the risks attached to sharing their personal data. In July 2019, there was a major global uproar around the use of an application called FaceApp, a fun online application that allowed users on Facebook to determine how they would look as they aged. This seemingly harmless application was developed in Russia and was accompanied by a set of terms and conditions which granted the developers full rights in perpetuity to all images uploaded to the application, for the developers to use as they deemed fit, in any way, at any time across the globe. On closer inspection, this does not seem to be a fair exchange between users and developers, which raises many questions related to cyber ethics and cybersecurity in the digital space.

A tangible benefit of using social media is put forward by Pavlov et al. (2018) who demonstrate how social media is used to calculate 'alternative metrics' (or altmetrics for short). Altmetric services track what impact a study has on social media and the traditional media, and in the case of the former is calculated based on mentions on Twitter, shares on Facebook and the number of people who read the stories on selected academic social networks. The OSINPI group tracked 15 articles and found that 'social media accounts clearly boost the metric scores of research group publications and the visibility of the research both within and beyond the scientific community' (Pavlov et al., 2018: 1). This resulted in multiple spin-offs for the scientists and researchers, including approaches from the traditional media to amplify the news and stories shared on social media.

There are other benefits to using social media including that of experiencing 'live science' as it unfolds and develops (Jepson, 2014). For example, in 2017 Professor Lee Berger and Professor John Hawkins, both associated with the University of the Witwatersrand, developed a live blog and Facebook page called The Daily Life of an Explorer[8] which enabled viewers from around the world to track, follow and engage in real time with scientists and explorers who were excavating for *Homo naledi*[9] in the Dinaledi Cave, part of the Rising Star Cave System located in the Cradle of Humankind just outside of Johannesburg, South Africa. The blog included images, live video coverage, podcasts, interviews with experts, scientists and explorers and opportunities for people to engage directly in real time with scientists on site. It also enabled users to share in the 'Eureka! Moments' when new hominid fossils were discovered or the tense moments when explorers found it difficult to squeeze through parts of the cave.

Another example of using new, creative media technologies to make science accessible to wider publics was the development of a free mobile application by Professor Berger and his team in conjunction with the Perot Museum of Nature and Science in Texas, USA, which allows users, scientists and the general public to enjoy a virtual experience of the Dinaledi cave system in six different languages, using cardboard 3D glasses.[10]

8 https://www.facebook.com/dailylifeofanexplorer/

9 http://www.wits.ac.za/news/latest-news/research-news/2015/2015-09/homo-naledi/a-new-species/

10 https://www.wits.ac.za/news/latest-news/general-news/2018/2018-10/

The barriers to using social media

Some scientists are reluctant to use social media to communicate science as they have little knowledge of social media, do not know how to use it, do not to have the time to engage on social media, or see it as a frivolous, unprofessional activity that is age-based (Collins et al., 2016; Gardner & McKee, 2019; Pavlov et al., 2018). Other reasons put forward for shunning social media include the insufficiency of the medium's scientific validity, an aversion to the content being shared, the risk of being exposed to wide audiences, the erosion of privacy online and a dislike for various social media formats, especially Twitter's microblogging format or Instagram's obsession with vanity pictures.

In South Africa and the rest of the developing world, one of the greatest barriers to using social media is the lack of access to data and fast internet services, particularly outside of urban areas.

Social media risks

'It is important to question who holds power in society, who takes the important decisions, who owns the basic resources, who is considered influential, who has the reputation to influence and change society, who is an opinion maker and who defines the dominant norms, rules and values' (Fuchs, 2014: 7). This speaks to both the traditional and social media as we determine how these media platforms benefit some, whilst disadvantaging others. Fuchs advocates for the need to develop a society that is universally beneficial to all.

The influence of players in the digital sphere was revealed in a study by Broniatowski et al. (2018) who sought to better understand how Twitter bots and trolls promote online health content. In the study, 'bots' are described as 'social media accounts that automate contention promotion', whilst 'trolls' are described as 'individuals who misrepresent their identities with the intention

---

wits-and-perot-museum-launch-virtual-reality-app-of-dinaledi-cave.html

of promoting online discord' (Broniatowski et al., 2018: 1).

The study found that through the amplification of anti-vaccine messages by false social media accounts, content around vaccinations was polluted, and public consensus on social media was eroded. Another strategy was to represent both sides of the debate, whilst inherently promoting one perspective, similar to strategies employed amongst certain political groups in national election campaigns in the United States[11] and in South Africa.[12]

Broniatowski et al. (2018: 1) explain how 'health-related misconceptions, misinformation and disinformation spread over social media pose a threat to public health'. As social media users are exposed to erroneous information about vaccines, they take time to digest the information or to explore it further, and in so doing delay in taking the required action to vaccinate, thereby putting themselves and entire populations at risk. The study analysed a set of 1 793 690 tweets collected over three years and included a qualitative study of a Twitter hashtag which deliberately politicised the issue and created dissonance in the Twitter sphere. The #VaccinateUS hashtag was traced to Russian troll accounts linked to a company associated with the Russian government that was known for influencing issues online.

The study found that about half of the tweets about vaccinations analysed contained anti-vaccine sentiments, and that people were more likely to trust information on the internet and in social media groups than to trust their own healthcare providers or public health experts. The study concluded that 'whilst bots spread malware and unsolicited content in the form of anti-vaccine messages, Russian trolls promoted discord' online (Broniatowski et al., 2018: 1).

Similar considerations should be given to science quackery, misinformation and disinformation campaigns on social media, some of which relate to climate change, medical conditions and

11 https://af.reuters.com/article/worldNews/idAFKCN1QF29E

12 https://www.dailymaverick.co.za/article/2019-03-06-beware-of-the-bots-and-the-trolls-in-the-polls/

even water quality in South Africa (Kubheka, 2017; Volmink, 2017).[13]

Another threat is posed through the social media conglomerates that own the large social media networking sites. A key finding from the *#SocialSA_2018* report by Patricios and Goldstuck (2018) reflects that influencers on social media can help recruit audiences and turn users into advocates for a course, but that these influencers are not necessarily celebrities. They also question how companies and organisations can win back audiences and communities grown and developed on major social media networks, and 'migrate' them back to in-house networks. This is because the larger social media networks now want organisations to pay to advertise to the communities that the organisations helped to build over time. Organisations have no control over the algorithms used by the major social networking platforms, no influence over the management of these platforms and no access to the data derived from the communities that they helped to develop on social media.

The Fourth Industrial Revolution brings with it increased interaction between humans and artificial intelligence (AI) in the form of machine learning, chat bots and content generated by AI, which may bring with it its own challenges, especially those related to issues of privacy, governance and ethics.[14]

Who is using social media?

Data published by the Pew Research Centre (2019) shows that 98% of young adults (i.e. 18–29 year olds) in the US use the internet, whilst 88% use social media. About half of all young

---

13 The Centre for Science and Technology and Mass Communication held a conference titled 'Quackery and Pseudoscience' in 2017. Several videos that speak to this statement are available: http://www.censcom.com/index.php/conferences/conference-videos. See also Statement by Professor Jimmy Volmink, Dean of the Faculty of Medicine and Health Sciences (FMHS) at Stellenbosch University (2017): https://www.sun.ac.za/english/Lists/news/DispForm.aspx?ID=5088

14 The 4IRSA partnership's Digital Economy Summit: www.4irsa.org

adults in the US obtain their news online, with 32% indicating that social media is a major news source (Gottfried & Shearer, 2016). Only 5% of young adults obtained their news from newspapers, 14% from radio and 27% from television (Mitchell et al., 2016).

Data released by the US's National Science Board (2018), indicates that 81% of young adults use the internet as their primary source of science and technology information while 83% use it as their primary source to learn about science and technology. The high use of the internet and digital channels to access science and technology news is corroborated by a study undertaken by Hargittai et al. (2018), which indicates that 96% of young adults in the US turn to the internet for information about science and research, and almost two-thirds do so weekly. The latter study goes one step further in trying to measure science engagement online. It determined that more than 80% of the young adults surveyed, clicked or commented on information related to science and research, with content related to health and fitness being extremely popular.

Social media penetration is the highest in North America with about 70% of the population connected, followed by Europe (54%–66%), Asia (64%), and southern Africa (31%) (Hootsuite, 2018). On the other hand, in terms of mobile connectivity, the number of mobile connections in southern Africa in relation to the population is 147%, which is way above the global average of 112% and North America (103%) (Hootsuite, 2018).

In terms of active users of key global social media platforms, Facebook is the largest networking site in the world with 2.1 billion users (Hootsuite, 2018). It is the easiest social network through which to reach mass markets, with steady engagement from fans and followers. However, it must be noted that the rate of engagement does not necessarily translate to influence or behavioural change (Sanne & Wiese, 2018). A literature review by Schein et al. (2011) also concluded that the impact of social media on the awareness of issues and behavioural change related to healthcare communication is still to be determined.

Like Facebook, LinkedIn can also be described as a social networking site that seeks to connect professionals, with 260 million users around the world in January 2018 (Hootsuite, 2018).

Social media platforms that are renowned for their visual and audio-visual content are Instagram and YouTube. Instagram, with 800 million users, is used to share visual material, especially images and short video clips, and enjoys a higher engagement rate than Facebook (Hootsuite, 2018). On the other hand, YouTube a video sharing site, has 1.5 billion subscribers globally (Hootsuite, 2018).

With almost 7 000 tweets being sent per second, Twitter is a microblog with 330 million users (Hootsuite, 2018) that is used to engage on events, for live tweeting and for sharing or amplifying news, politics and related content, with limited engagement.

Finally, blogs are usually theme or topic-specific mini websites like BlogSpot or Wordpress, where people can create and publish lengthier pieces. See section on alternative media below, for more information on platforms like *Medium* which can be described as a 'blog for blogs'.

The South African landscape

With a population of about 57 million, South Africa has a 54% internet penetration rate which amounts to about 31 million internet users (January 2018 data), up 7% from 2017 (Hootsuite, 2018). There are 18 million active social media users in the country, with a 20% year-on-year increase in subscriptions to various social media platforms (Hootsuite, 2018). There are 38 million unique mobile users in South Africa, of which 16 million are active social mobile users. In terms of device usage, 95% of South Africans have mobile phones, of which 60% are smartphones (Hootsuite, 2018).

Facebook is the most popular social media platform in South Africa, in use by 46% of the South African population (about 18 million users, of which about 14 million access Facebook via mobile devices) (Hootsuite, 2018). The introduction of Facebook Lite, which has been adapted for the South African context, has made it easier for people to access this social medium in recent years (Patricios & Goldstuck, 2018). This is also the most popular social media platform for advertising and rivals the traditional broadcast media to reach a broader audience. Communication is two-way or multi-way and can be measured precisely.

There are about 8 million South Africans on Twitter and this is the application that generates the most debate on politics, the economy, hard news and crime. It boasts strong user engagement and is the social medium of choice that facilitates communication, engagement and public discourse (Hootsuite, 2018).

Instagram is growing steadily in South Africa with just over a million users and is the social medium most used by young people (Hootsuite, 2018). There are about 6.1 million South African LinkedIn users. It must be noted that there have been dramatic declines in the use of Pinterest, Google+, WeChat, WhatsApp and Snapchat in the country (Patricios & Goldstuck, 2018).

Integrating traditional and social media

Developing an integrated science communications strategy

Scientists are adopting blended approaches and are combining the use of traditional and social media to make their research visible. For example, a scientist who has published a research paper will create a mini-website online as a hub on which to host the academic paper (or links to the journal), a media release, fact sheets, images, video material, podcasts, captions, background information on the scientist and collaborators, and other information, thus making use of owned media channels to host the news.[15] These materials can then be shared via social media or a link to cloud-based file storage services to reach out to and to create awareness of the research amongst science journalists, fellow scientists, the general public and other social actors.

At the same time, a traditional media advisory can be shared

15 For an example of a basic microsite, visit https://www.wits.ac.za/homonaledi/

with science journalists, news wires and online press offices like the American Association for the Advancement of Science's Eurekalert! site. Scientists can write an opinion piece for a weekly newspaper like the *Mail & Guardian* in South Africa, which has a small circulation but a quality audience with thoughtful readers who will engage with the subject matter, or the *Sunday Times* which, with a circulation of 240 219 and a potential readership of 1.2 million, remains a high-impact publication in South Africa (Breitenbach, 2019). This is the same for traditional radio and television interviews, specialist documentary programmes or in-depth news features on news programmes like *Carte Blanche*. So, how are integrated science communications strategies developed and executed?

Six key elements should be considered when planning an integrated science communication and engagement strategy. It is important to determine the aim, objectives or goal of the communication; to identify the audiences with whom to engage; to develop the messages to be conveyed; to consider the medium to be used as a conduit to reach select audiences; to agree on the science communication and engagement activities to be implemented; and to identify upfront how to monitor and evaluate the communications activities.

These elements are as important when considering integrated science communications that include digital media (online, new and social media), which include all the elements described above, although with a different emphasis on some aspects rather than others.

Goals and objectives

It is essential for the aim, objectives and goals of an integrated science communication plan to be established upfront, as it speaks to what is to be achieved, how it is to be achieved and by when it is to be accomplished. It provides direction for how messages are developed, which media are to be used, which channels and platforms are best suited for use, and which science communication and engagement activities are implemented.

Audiences, publics and communities

As in traditional science communication, it is important to understand with whom it is important to communicate, and to understand the demographics, geographies and psychographics of the potential audiences. In the digital sphere, through access to rich data, tracking and analysis, audiences can be highly segmented based on their digital footprints, and with the necessary permissions, their content preferences.

The ethics involved in data mining, collection and analysis, must be considered in line with the consent of users who should first grant access to their respective online identities and digital footprints. At the same time, social media allows for the growth and development of like-minded communities to share relevant information, to increase engagement, to develop collegiality and to congregate on topics and themes of shared interest.

Messages and content

The ability to tailor research messages and science content to individuals online or via specific social media platforms is effective and powerful, based on the digital profile of users. It differs from the key messages used in the traditional mass media where one set of messages can be shared across print and broadcast platforms. For social media, content per user or user group must be tailored for each online, digital or social media channel selected for publication.

There is also a need to generate multimedia content that can be adapted for use across different platforms. The content must be newsworthy, topical, should tell a story and, where possible, should include creative multimedia material that can be adapted across all platforms. The development of multimedia content allows for the ability to show and tell a research story, to interact with users, and to use visual content to create impact.

However, fundamental differences exist in how the science is communicated, how engagement occurs, the level or depth of interaction with the content and other social media users, the immediacy of the interaction, and the ability of the science communication or engagement activities to effect change. For example, whilst traditional media allows for feedback through letters to the editor or calls on radio talk shows, there is very limited engagement on issues related to science that features in the traditional media. This engagement is often staggered in newspapers, with feedback often only published if it is a major breakthrough, a new discovery or if it impacts on a large populace.

The medium matters

Scientists now have the ability to consider the traditional media as a conduit to make their knowledge accessible in the public sphere, and to combine it with a proliferation of social media networking tools. Researchers are able to use creative multimedia strategies to create their own content, to build their own communities and to communicate directly with audiences, fellow researchers and the general public in a personable way, across platforms. Scientists and professional science communicators are increasingly becoming digitally savvy.

Science communication tools and activities

The digital tools available to scientists and science communicators can largely be categorised using a model developed by Dietrich (2014) called the PESO model. The model encompasses four categories: paid, earned, social and owned media.

Paid media refers to digital and online advertising, native advertising, paid for or promoted digital content like sponsored tweets, Facebook or LinkedIn posts. The earned component speaks to the traditional public relations model in the digital sphere and includes media and influencer relations. The social media channels comprise of the myriad of available channels that can be used to share science and to make science accessible in the public sphere, with the most popular current social media platforms being Facebook, LinkedIn, YouTube, Twitter and Instagram. Finally, the owned media channels are those which are owned by research institutions and universities like websites, newsletters and social media channels, which are used to host or distribute content that is crafted and developed in-house and tailored for each audience and channel.

The ability to tailor science content and to adapt it to a variety of audiences across multiple platforms is a major benefit for scientists and science communicators, affording them direct control over the initial content, guaranteeing publication and distribution, and allowing for the ability to track users and content through analytics and other software tools that provide insight.

Monitoring and evaluation

It is easier to monitor and evaluate quantitatively digital and online campaigns through media monitoring agencies, Google analytics, online dashboards, social media reporting and online analysis tools. The complexity of social data also needs to be appreciated and disaggregated so as to develop user insights rather than just metrics. However, the way in which social media is evaluated using current metrics may soon be outdated, with organisations now looking to link the outcomes of social media campaigns to tangible benefits like increased collaboration amongst peers, lower promotion costs, increased funding and talent attraction (Patricios & Goldstuck, 2018).

Whilst qualitative monitoring and evaluation provides detailed insight into the success of campaigns or projects, it remains expensive.

Alternative media models

As scientists struggle with traditional media to publish their science news, alternative media models have developed over the years to make research accessible to a range of publics. *The Conversation*[16] is one such example, which is an independent source of news from the academic and research community, delivered directly to the public. It develops content in partnership with experts and researchers, usually attached to a university or research institution (which sometimes funds the platform) and then publishes and syndicates that content across its own platforms but also shares it with other traditional and online media.

There are benefits of these forms of media, including that the publications can produce quality, multimedia science news in partnership with academics and researchers, thereby creating a safe space for academics and researchers to engage and to jointly produce accurate content that is compelling and engaging to popular audiences. It also provides an opportunity for scientists to sign off their work before it is published, thereby reducing the traditional tension between scientists and journalists.

Issues related to the independence of these alternative media remain, particularly in terms of funding from universities, science agencies and the like, but similar tensions have been identified in the form of influence from advertisers and owners in the past.

A similar model is *Quartz.com*[17] founded in 2012 by journalists, a platform which carries science and technology news that is 'creative and intelligent journalism' told through stories and tailored for readers. This platform has subsequently been funded by the corporate sector.

Medium[18] is another example of online social journalism, which includes a mixture of people, publications and blogs and which is best described as a 'blog host' that offers quality content and pays authors depending on the number of people who engage with the content. *Medium* is funded through subscriptions, native advertising and the sponsorship of some article series.

---

16 www.theconversation.com/Africa

17 https://qz.com

18 https://medium.com/

Future disruption

The advent of the Fourth Industrial Revolution is set to dramatically change how humans interact with technology, how we express ourselves, how we communicate and how we engage with Society 5.0.

Whilst digitalisation has already transformed newsrooms and the way in which science is communicated, the potential for further disruption through transformative technologies such as artificial intelligence, big data, automation and crypto currencies is immense. These new technologies have already, and will continue to impact our lives and the world of work as we know it today – it will reshape how we live, work, and interact with each other.[19]

For example, algorithms and artificial intelligence are already determining online advertising in publications, writing and curating digital content, and guiding online promotions. Native advertising and product placement are now automated across digital spaces, which pose a risk to digital communication. Future science content on social media is likely to include rich content, coupled with podcasts, infographics, animation, footage from drones and Go Pros, in real time. Access to open and live science is also in demand.

Science communication lends itself to interactive mobile applications and virtual, augmented and mixed reality experiences, which are already in existence. The use of edutainment, gaming, and reality shows are on the cards, along with new, interactive platforms.

At the same time, given the risks associated with these new content types, innovative creative media technologies, and multiple platforms, some of which are known, and the majority of which are still to be realised, there is a need for the development of ethical guidelines to combat bots, trolls and potential 'weaponised' communication. This includes the development of new privacy laws to protect digital footprints and social media users, as well as

19 www.4irsa.org

policies to distinguish fact from fiction, and applications to distinguish science from quackery, and scientists from bots.

Conclusion

The Fourth Industrial Revolution brings with it the opportunity to use digital technologies that will enable the creation of innovative media platforms, the ability to develop rich, science-based creative content and the confidence to share it with multiple publics at a lower cost than through the use of traditional media. At the same time, the contextual realities of South Africa and the continent where the future of work is uncertain, where inequality remains rife and where the digital divide has the potential to further isolate communities who have little or no access to data or the internet, must be acknowledged.

The various pressures placed on scientists to make research and knowledge accessible emanate from various social actors, including funders. There are several case studies that demonstrate the benefits of communicating science, with one of the key advantages being the ability to empower citizens by having them participate in science.

At the same time, the changing media environment and fluctuating news values in a digital world dictate that the traditional media no longer serves as the sole conduit of science to the general public. This makes it necessary for scientists to find alternative ways of communicating to multiple publics, despite the tensions that may exist between scientists and the media. Safer options include using hybrid media outlets like *The Conversation* that encourages academics and journalists to partner to develop factual news that is accessible to lay publics. Other choices include the use of academic social media like ResearchGate, as well as closed Facebook and WhatsApp groups that are shared by like-minded scientists, issue publics or persons interested in a particular research topic, theme or study.

There is a steady uptake in the use of social media to communicate science across the world, including in South Africa, where an integrated approach to science communication is proposed. This includes the use of the traditional media and social media to make science visible in the public sphere. There are many case studies both locally and abroad of how integrated science communication has succeeded through the consolidation of paid, earned, social and owned media.

However, some scientists remain reluctant to use social media as they do not understand the platforms, how they work or are sceptical about their scientific validity. On the other hand, some of the benefits put forward include bridging the gap between science and society, empowering citizens in real time, demonstrating social impact and contributing to democracy.

Many social media risks are linked to media ownership, power and politics in society, which pose risks to science, scientists and social media users. The few conglomerates that dominate social media like Facebook and Twitter are powerful entities that have access to the private data of millions of users. This brings skewed power relations between media conglomerates and social media users, especially with regard to risks related to governance, privacy and ethics in the digital sphere. The use of artificial intelligence and machine learning driven by algorithms have the ability to power bots and trolls, to influence the content to which users are exposed, to sell data to advertisers, to manipulate how users think, and what they think about, amongst other risks.

There are endless opportunities to use new creative multimedia technologies to facilitate science communication across multiple platforms in real time across physical and virtual boundaries. However, there are concomitant risks to science and science engagement, some of which are known, and others which we can only predict.

References

Allan, S. (2009). The future of science journalism. *Journalism,* 10(3), 280–282. doi: 10.1177/1464884909102570

Badenschier, F. & Wormer, H. (2012). Issue selection in science journalism: Towards a special theory of news values for science news? In S. Rödder, M. Franzen & P. Weingart (eds), *The Sciences' Media Connection: Public communication and its repercussions* (pp. 59–85). Sociology of the Sciences Yearbook 28. Dordrecht: Springer.

Berlin Declaration on Open Access to Knowledge in the Sciences and Humanities (2003). https://openaccess.mpg.de/Berlin-Declaration

Bik, H. & Goldstein, M. (2013). An introduction to social media for scientists. *PLoS Biology*, 11(4), e1001535.

Bourguignon, J.-P. (2018, 16 September). Scientists can lead the fight against fake news. World Economic Forum website. https://www.weforum.org/agenda/2018/09/scientists-can-lead-the-fight-against-fake-news/

Breintenbach, D. (2019, 13 August). Newspapers ABC Q2 2019: Newspapers' declining circulation trend continues. *Bizcommunity*. https://www.bizcommunity.com/Article/196/90/194166.html

Breytenbach, K. (2005). Media moving to 'juniorisation of newsrooms'. Interview with Joe Thloloe, Chairperson of the South African National Editors' Forum. *Independent Online*, May. https://www.iol.co.za/news/politics/media-moving-to-juniorisation-of-newsrooms-240348

Broniatowski, D., Jamison, A., Qi, S.H., AlKulaib, L., Chen, T., Benton, A., et al. (2018). Weaponized health communication: Twitter bots and Russian trolls amplify the vaccine debate. *American Journal of Public Health*, 108(10), 1378–1384.

Brossard, D. (2013). New media landscapes and the science information consumer. *Proceedings of the National Academy of Sciences of the United States of America*, 110(3), 14096–14101.

Bucchi, M. & Neresini F. (2008). Science and public participation. In E. J. Hackett, O. Amsterdamska, M. Lynch & J. Wajcman (eds), *The Handbook of Science and Technology Studies* (pp. 449–472). Cambridge, MA: MIT Press.

Bucchi, M. & Mazzolini, R. (2003). Big science, little news: Science coverage in the Italian Daily Press, 1946–1997. *Public Understanding of Science,* 12(1), 7–24.

Chung, I.J. (2011). Social amplification of risk in the Internet environment. *Risk Analysis,* 31, 12.

Claassen, G. (2011). Science and the media in South Africa: Reflecting a 'dirty mirror'. *Communicatio,* 37(3), 351–366.

Collins, K., Shiffman, D. & Rock, J. (2016). How are scientists using social media in the workplace? *PLoS One,* 11(10), e0162680. https://doi.org/10.1371/journal.pone.0162680.

Cornia, A., Sehl, A. & Nielsen, R. K. (2018). 'We no longer live in a time of separation': A comparative analysis of how editorial and commercial integration became a norm. *Journalism*. London: Sage.

Daniels, G. (2018a, 10 July). Journalist bloodbath: New beat research to find out what happens to journos and journalism. *The Media Online*.

Daniels, G. (2018b). In A. Finlay (ed.), *State of the Newsroom Report 2018: Structured/unstructured*. Johannesburg: Wits Journalism. https://journalism.co.za/ wp-content/uploads/2019/07/State-of-the-Newsroom-report-2018_updated-20190709.pdf

Dahlgren, P. (2009). *Media and Political Engagement: Citizens, communication and democracy*. New York: Cambridge University Press.

Dean, C. (2002) New complications in reporting on science: Scientists have important roles to play in getting the news right, but they are often reluctant participants. *Questia*. https://www.questia.com/magazine/1G1-93534166/new-complications-in-reporting-on-science-scientists

De Lanerolle, I. (2016). The message is the medium: fluid public-making in South African social media. Unpublished document. University of the Witwatersrand, Johannesburg.

Department of Science And Technology (DST) (2019). White Paper on Science, Technology and Innovation 2019. https://www.dst.gov.za/images/2019/White_paper_web_copyv1.pdf

Department of Science And Technology (DST) (2017). Department of Science and Technology South Africa Annual Report 2017. Pretoria. http://www.dst.gov.za/images/DST-AR-2017-WEB.pdf

Dietrich, G. (2014). *Spin Sucks: Communication and Reputation Management in the Digital Age*. Indianapolis, IN: Que.

Edelman Trust (2018). The 2018 Edelman Trust Barometer Global Report. Chicago: Edelman, http://cms.edelman.com/sites/default/files/2018-2/2018_Edelman_Trust_Barometer_Global_Report_FEB.pdf

Finlay, A. (ed.) (2017). *State of the Newsroom 2017: Fakers and makers*. Johannesburg: Wits Journalism. http://wits.journalism.co.za/wp-content/uploads/2018/03/906-STATE-OF-THE-NEWSROOM-2018-REPRINT-V2.pdf

Finlay, A. (2018). *The State of the Newsroom Report*. Johannesburg: Wits Journalism.

Flanagin, A. J., Hocevar, K. P. & Samahito, S. N. (2014). Connecting with the user-generated web: How group identification impacts online information sharing and evaluation. *Information, Communication and Society,* 17(6), 683–694. doi: 10.1080/1369118X.2013.808361.

Franklin, B. (2004). *Packaging Politics. Political communications in Britain's media democracy.* London: Arnold.

Fuchs, C. (2014). *Social Media: A critical introduction*. London: Sage.

Galtung, J. & Ruge, M. H. (1965). The structure of foreign news. *Journal of Peace Research,* 2(1), 64–91.

Gardner, J. M. & McKee, P. H. (2019). Social media use for pathologists of all ages. *Archives of Pathology and Laboratory Medicine,* 143(3), 282–286.

Gascoigne, T. & Metcalfe, J. (2012). *Planning Communication into Science: Deliver impact from your research*. Econnect Communication: Australia.

Gastrow, M. (2015). Science and the social media in an African context: The case of the Square Kilometre Array Telescope. *Science Communication,* 37(6), 703–722.

Gottfried, J. & Shearer, E. (2016). *News Use across Social Media Platforms 2016*. Pew Research Center. Washington DC.

Gumede, W. (2014). South Africa's media and the strengthening of democracy. *Rhodes Journalism Review,* 34, 7–10.

Hamshaw, R., Barnett, J. & Lucas, J. (2017). Framing the debate and taking positions on food allergen legislation: The 100 chefs incident on social media. *Health, Risk and Society,* 19(3–4), 145–167.

Harcup, T. & O'Neill, D. (2001). What is news? Galtung and Ruge revisited. *Journalism Studies,* 2(2), 261–280.
Harcup, T. & O'Neill, D. (2016). What is news? *Journalism Studies,* 18(12), 1470–1488.
Hargittai, E., Fuchslin, T. & Schafer, M. (2018). How do young adults engage with science and research on social media? Some preliminary findings and an agenda for future research. *Social Media and Society,* 1–10.
Hart, R. & Shaw, D. (2001). *Communication in US Elections: New agendas.* Lanham: Rowman and Littlefield.
Hootsuite (2018). We Are Social 2018 Global Digital Report. https://digitalreport.wearesocial.com/.
Hwong, Y.L., Oliver, C., Van Kranendonk, M., Sammut, C. & Seroussi, Y. (2017). What makes you tick? The psychology of social media engagement in space science communication. *Computers in Human Behavior,* 68, 480–492.
ICASA (2019). ICASA statement on Competition Commission report. https://www.icasa.org.za/news/2019/icasa-welcomes-the-provisional-findings-and-recommendations-by-the-competition-commission
Jasanoff, S. (ed.) (2004). *States of Knowledge: The co-production of science and the social order.* London: Routledge.
Jha, A., Lin, L. & Savoia, E. (2016). The use of social media by state health departments in the US: Analyzing health communication through Facebook. *Journal of Community Health,* 41, 174–179.
Jepson, K. M. (2014). An examination of the effects of digital media on the communication of science. Unpublished PhD thesis, Montana State University.
Jones, D. (2004). Why Americans don't trust the media: A preliminary analysis. *Harvard International Journal of Press/Politics,* 9(2).
Joubert, M. & Guenther, L. (2017). In the footsteps of Einstein, Sagan and Barnard: Identifying South Africa's most visible scientists. *South African Journal of Science,* 113(11–12), 1–9.
Khumalo, B & Minors, D. (n.d.) HIV Liver Transplant Social Media. Unpublished report prepared for the University of the Witwatersrand.
Kiernan, V. (2006). *Embargoed Science.* Urbana: University of Illinois Press.
Kizer, Z. (2018). Not a scientist: how politicians mistake, misrepresent, and utterly mangle science. *Journal of Science Communication,* 17(1). https://doi.org/10.22323/2.17010704
Kruger, F. (2018). Preface. In A. Finlay (ed.), *State of the Newsroom Report 2018: Structured/unstructured.* Johannesburg: Wits Journalism. https://journalism.co.za/wp-content/uploads/2019/07/State-of-the-Newsroom-report-2018_updated-20190709.pdf
Kruger, F. (2017). Preface. In A. Finlay (ed.), *State of the Newsroom Report 2017: Fakers and makers.* Johannesburg: Wits Journalism. http://www.journalism.co.za/wp-content/uploads/2018/03/WITS-STATE-OF-THE-NEWSROOM_March_2018.pdf
Kubheka, B. (2017). Ethical and legal perspectives on the medical practitioners' use of social media. *South African Medical Journal,* 107(5), 386–389.

Kuttschreuter, M., Rutsaert, P., Hilverda, F., Regan, Á., Barnett, J. & Verbeke, W. (2014). Seeking information about food-related risks: The contribution of social media. *Food Quality and Preference,* 37, 10–18.

Lynch, J. & Condit, C. (2006). Genes and race in the news: A test of competing theories of news coverage. *American Journal of Health Behaviour,* 30(2), 125–135.

Mitchell, A., Gottfried, J., Barthel, M. & Shearer, E. (2016). *The Modern News Consumer: News attitudes and practices in the digital era.* Washington, DC: Pew Research Center.

Muchendo, C. (2005). Sourcing of HIV/AIDS treatment news: A case study of selected South African print media. Unpublished master's thesis, University of the Witwatersrand.

Mudde, S. (2019). A tale of two citizens: How South Africa's most visible scientists use Twitter to communicate with the public. Unpublished master's thesis, Stellenbosch University.

National Research Foundation. (2018). National Research Foundation's Science Engagement Strategy. Pretoria. https://www.nrf.ac.za/science-engagement

Nelkin, D. (1995). *Selling Science: How the press covers science and technology.* New York, NY: W.H. Freeman.

O'Neill, D. & Harcup, T. (2009). News values and selectivity. In K. Wahl-Jorgensen & T. Hanitzsch (eds), *The Handbook of Journalism Studies* (pp. 161–175). London: Routledge.

Onyancha, O. (2015). Social media and research: An assessment of the coverage of South African universities in ResearchGate, Web of Science and the Webometrics Ranking of World Universities. *South African Journal of Libraries and Information Science,* 81(1), 8–20.

Palitza, K., Ridgard, N., Struthers, H. & Harber, A. (eds) (2010). *What is Left Unsaid: Reporting the South African HIV Epidemic.* Johannesburg: Jacana.

Patel, S. (2019). How does science become news in South Africa? A quantitative and qualitative analysis of the news values and factors that influence the publication of science in South African newspapers .Unpublished master's, University of the Witwatersrand.

Patricios, O. & Goldstuck, A. (2018). At a glance: The South African social bedia Landscape 2018. *#SocialSA_2018.* Ornico, http://website.ornico.co.za/wp-content/uploads/2017/10/SML2018_Executive-Summary.pdf.

Pavlov, A. K., Meyer, A., Rosel, A., Cohen, L., King, J., Itkin, P. et al. (2018). Does your lab use social media? Sharing three years of experience in science communication. *Bulletin of the American Meteorological Society.*

Peters, H. P. (2013). Gap between science and media revisited: Scientists as public communicators. *Proceedings of the National Academy of Sciences,* 3, 14102–14109.

Pew Research Centre (2019) Internet/broadband fact sheet. https://www.pewresearch.org/internet/fact-sheet/internet-broadband/

Rudat, A. & Budar, J. (2015). Making retweeting social: The influence of content and context information on sharing news in Twitter. *Computers in Human Behavior.* doi: 10.1016/j.chb.2015.01.005

Sanne, P. & Wiese, M. (2018). The theory of planned behaviour and user engagement

applied to Facebook advertising. *South African Journal of Information Management,* 20(1). https://doi.org/10.4102/sajim.v20i1.915.
Schein, R.. Wilson, K. & Keelan, J. (2011). Literature review on the effectiveness of the use of social media: A report for PEEL Public Health, Canada. https://www.peelregion.ca/health/resources/pdf/socialmedia.pdf
Schudson, M. (2003). *The Sociology of News.* New York: WW Norton.
Searle, J. (1979). *Expression and Meaning: Studies in the theory of speech acts.* London: Cambridge University Press.
Stacey, N., Tugendhaft, A. & Hofman, K. (2017). Sugary beverage taxation in South Africa: Household expenditure, demand system elasticities, and policy implications. *Preventative Medicine,* 105S, S26-S31. doi: 10.1016/j.ypmed.2017.05.026. Epub 2017 Jun 1.
Stromback, J., Karlsson, M. & Hopmann, D. (2012). Determinants of news content: Comparing journalists' perceptions of the normative and actual impact of different event properties when deciding what's news. *Journalism Studies,* 13(5–6), 718–728.
Stromback, J. & Karlsson, M. (2011). Who's got the power? Journalists' perceptions of changing influences over the news. 2011. *Journalism Practice,* 5(6), 643–656.
Thloloe, J. (2005, 5 May). Media moving to juniorisation of newsrooms. *The Star.*
Van Rooyen, C. (2002). A report on science and technology coverage in the SA print media. Unpublished master's thesis, Stellenbosch University.
Volmink, J. (2017). Summit to counter quackery, pseudoscience and fake news in healthcare. Stellenbosch University website. https://www.sun.ac.za/english/lists/news/DispForm.aspx?ID=5088
Weingart, P. (2001). *Die Stunde der Wahrheit? Zum Verhältnis der Wissenschaft zu Politik, Wirtschaft und Medien* (The moment of truth? On science's relationship with politics, economics and the media in the knowledge society). Weilerswist: Velbrück.
Wellcome Trust (n.d.). Public Engagement Fund. Wellcome Trust UK website. https://wellcome.ac.uk/funding/schemes/public-engagement-fund
Wilcox, S. (2003). Cultural context and the conventions of science journalism: Drama and contradiction in media coverage of biological ideas about sexuality. *Critical Studies in Media Communication,* 20(3), 225–247.
World Economic Forum (WEF) (n.d.). Internet for all. World Economic Forum website. https://www.weforum.org/projects/internet-for-all
Yeo, S. (2016). What science says about key considerations for communication. Presentation at the University of Utah. http://nas-sites.org/emergingscience/files/2016/09/Yeo.pdf
Yeo, S., Binder, A., Dahlstrom, M. & Brossard, D. (2018). An inconvenient source? Attributes of science documentaries and their effects on information-related behavioral intentions. *Journal of Science Communication,* 17(2). https://doi.org/10.22323/2.17020207.
Yeo, S. K., Liang, X., Brossard, D., Rose, K. M., Korzekwa, K., Scheufele, D. A. & Xenos, M. A. (2017). The case of #arseniclife: Blogs and Twitter in informal peer review. *Public Understanding of Science,* 26, 937–952.

7 The quackery virus: An overview of pseudoscientific health messages on Twitter

George Claassen

Introduction

Quackery, the spread of pseudoscience and alternative 'facts' in the health field, has become a growing concern and challenge not only to scientists but also to journalists and the community at large. Trying to counter and refute misleading, harmful and often fraudulent health messages and marketing in the age of social media has been the source of various studies as well as intensive scrutiny of specifically the phenomenon of Twitter as a medium through which these messages are spread (Anderson et al., 2010; Bik & Goldstein, 2013; Birch, 2011; Colditz et al., 2018; Joubert & Costas, 2019; Mudde, 2019; Nagy et al., 2018; Steinman, 2018; Su et al., 2017; Van Rooyen, 2017).

Furthermore, the public understanding of science is vital in any society, not only to counter fake news and pseudoscientific claims and quackery, but also to assist in finding ways to bring sound and trustworthy scientific findings to the attention of uninformed and often ignorant citizens bombarded by social media (Bauer, 2000; British Royal Society, 1995, 2007; Bucchi, 2004; Claassen, 2011; Gastrow, 2015; Hartz & Chappell, 1997; Hesmondhalgh, 2008; Joubert, 2018; Joubert & Guenther, 2017; Mooney & Kirshenbaum, 2009; Nelkin, 1995; Shukla & Bauer,

2007; Steinman, 2014; Webster, 2006). Bauer (2008: 119) states that this brings new emphasis to the reasons why science should be important to society and the public, as it is 'important for making informed consumer choices'. Applied to the health field where quackery and pseudoscience can cause serious harm, this becomes even more imperative. Yet the way the media are used by consumers, have changed drastically over the past decade, as Bell points out (2016: n.p.):

> Our news ecosystem has changed more dramatically in the past five years than perhaps at any time in the past five hundred [...] *Social media hasn't just swallowed journalism, it has swallowed everything*' (emphasis added).

This overview analyses how social media, and specifically Twitter, is a social media environment and platform where false health claims and quackery often spread virally and become part of what Habermas (1991: 30) describes as the public sphere, now vastly expanded from his 'coffee houses, […] salons and […] table societies' to the highly active viral environment of false health messages propagated on social media and specifically, for the purposes on this study, Twitter.

Identifying quackery and pseudoscience, what Pigliucci calls 'nonsense on stilts' (2010), and distinguishing it from valid evidence-based science, has become one of the most important science communication challenges over the past decade, but it also is an endeavour in morality. When people and specifically celebrities who are followed by millions on social media platforms by an often gullible public, make false claims on health, it becomes not only imperative that scientists and informed journalists counter their non-evidence-based claims (Franzen et al., 2007; Hall, 2014), but it also becomes a moral issue (Claassen, 2019a; Pigliucci, 2010; Pigliucci & Boudry, 2013). Quoting the 19th century British scientist Thomas Henry Huxley on the moral duty of everyone in society to make a distinction between science and non-science, Pigliucci (2010: 1) touches upon an often-neglected reason why science communication is so important in society:

> The foundation of morality is to [...] give up pretending to believe that for which there is no evidence, and repeating unintelligible propositions about this beyond the possibilities of knowledge.

Pigliucci (2010: 1) goes further, emphasising how the dangers inherent in accepting pseudoscience and quackery can harm, and that to accept 'pseudoscientific untruths or conversely rejecting scientific truths, has consequences for us all, psychological, financial, and in terms of quality of life. Indeed [...] pseudoscience can literally kill people'.

Pigliucci and Boudry (2013) warn about the difficulty of distinguishing science from pseudoscience or non-science, and of demarcating the fields clearly, one of the reasons quackery can spread so quickly like a virus on Twitter. This results in a public that cannot make a clear distinction between evidence-based science, and quackery and fraudulent pseudoscientific claims. Bucchi (1998: 17) calls this distinction a 'demarcation between orthodoxy (science) and deviance (non-science)'. Pigliucci and Boudry question the demarcation problem and Laudan's (1983) premise about its demise and that it does not exist. Quacks and pseudoscientists are 'master mimics' at dressing their pseudoscientific claims in a scientific cloak, fooling and confusing the public, and it remains an ever-growing challenge for lay people to make sense of the validity of claims (Claassen, 2019a: 202–203; Pigliucci & Boudry, 2013).

Bronner (2011: 2) refers to a paradox between the 'coexistence of progress in human knowledge with the persistence of certain ideas that are either false or questionable' as conspiracy theories about the moon landings, the terrorist attacks of 11 September 2001 and, in the health field, homoeopathy and other pseudoscientific practices illustrate (Gray, 2019).

Ruse (2013) analyses the Gaia Hypothesis and why it was so strongly rejected by scientists, mostly evolutionary biologists, but widely accepted by members of the public, also on Twitter, as another 'illustration of the intermixed world – although most of the time because of a lack of knowledge about the scientific methods and evidence-based science among the public – of science and pseudoscience' (as cited by Claassen, 2019a: 203).

The health risk caused by the way news is reported is also emphasised by Nelkin (1995: 47), and science and specifically health-related journalism are often criticised for 'inaccurate or misleading' reporting (Dentzer, 2009: 1). This danger is seriously enlarged by false health claims made on Twitter.

How to counter this? Studies have shown that it is indeed beneficial for scientists and scientific institutions to have an online social media presence (Bik & Goldstein, 2013; Joubert, 2018; Joubert & Costas, 2019; Mudde, 2019; Van Rooyen, 2017), that Twitter can foster better public engagement with science (Jarreau, 2016), and that communicating science on Twitter 'works partly by relaying science to a more diverse audience' (Novak, 2015). They have also shown that social media users are often intermediaries making visible and assessing evidence of social impact (Pulido et al., 2018).

As one study points out, 'Scientists are increasingly using Twitter as a tool for communicating science. Twitter can promote scholarly discussion, disseminate research rapidly, and extend and diversify the scope of audiences reached. However, scientists also caution that if Twitter does not accurately convey science due to the inherent brevity of this media, misinformation could cascade quickly through social media' (Bombaci et al., 2016: 216).

The presence of scientists on Twitter and their role in making science news more quickly available and more accessible is leading to serious and often vigorous debate in the scientific community, especially after the publication of the 'Kardashian Index' (or K-Index) by Hall (2014). He defines the K-Index as a 'measure of discrepancy between a scientist's social media profile and publication record based on the direct comparison of numbers of citations and Twitter followers' (Hall, 2014: 1).

Hall (2014: 1) goes further, emphasising that in 'the age of social media there are people who have high-profile scientific blogs or twitter feeds but have not actually published many peer-reviewed papers of significance; in essence, scientists who are seen as leaders in their field simply because of their notoriety'.

Yet Twitter brings an important advantage to scientists and science in general to announce their or other scientists' studies and findings, breaking down the pay-wall problem for the general public of accessibility of scientific research by posting a link to the peer-reviewed article. This also assists journalists following Twitter to have quick and easy access to peer-reviewed studies which were before the age of social media often hidden behind expensive pay-walls.

Twitter as choice by scientists in science communication

Twitter is a special case with regard to science communication utilised by scientists themselves, as pointed out by various studies (Allgaier et al., 2013; Collins et al., 2016; Joubert, 2018; Vraga & Bode, 2017; Yeo et al., 2014), and as emphasised by Joubert and Costas (2019: 2):

> Twitter, a microblogging platform and social networking tool, has emerged as a particularly popular and powerful science communication platform that researchers (or scholars) embrace more readily than other social media platforms, possibly because it is viewed as more professional and more suitable for science communication than other tools such as Facebook.

There is reason for this popularity of scientists utilising social media and specifically Twitter to discuss and announce scientific developments. Traditional 'news publishers have lost control over distribution of news [...] Now the news is filtered through algorithms and platforms which are opaque and unpredictable', as Bell points out (2016). The 'inevitable outcome of this is the increase in power of social media companies. The largest of the platform and social media companies, Google, Apple, Facebook, Amazon, and even second order companies such as Twitter,

Snapchat and emerging messaging app companies, have become extremely powerful in terms of controlling who publishes what to whom, and how that publication is monetized' (Bell, 2016: n.p.).

Due to its immediacy in spreading news, Twitter has also become a very popular social media platform for scientists to counter pseudoscience and quackery and specifically unscientific, non-proven claims made by celebrities in support of misleading and often even fraudulent marketing of products and ideas by non-scientists. In the health field, this is rife (Ernst & Smith, 2018; Jarvis, 1997; Singh & Ernst, 2018; West, 2019).

Twitter has also become a popular vehicle to spread pseudoscience, and charlatans and quacks, often very influential when they are also celebrities, use it with great success to spread fake claims and news about health issues. A prime example is the actress Gwyneth Paltrow whose venture into the field of quackery through her goop website (Twitter handle @GwynethPaltrow or @goop, the latter described as 'The latest from goop, a lifestyle publication and shop'), has led to immense criticism from medical scientists and health practitioners. Examples of just a few scientists countering quackery on Twitter are medical specialist Jennifer Gunter (@DrJenGunter), health law and science policy author of *Is Gwyneth Paltrow Wrong About Everything?* Timothy Caulfield (@CaulfieldTim), medical scientists David Gorski (@gorskon) and Alastair McAlpine (@AlastairMcA30), a trauma surgeon under the pseudonym @DocBastard, Edzard Ernst, co-author of *Trick or Treatment: Alternative medicine on trial* Edzard Ernst (@EdzardErnst), award-winning scientist and science author Simon Singh who famously won a case against the British Chiropractic Association when sued for defamation (@SLSingh) (Boseley, 2009), and the health activist and project director of the Good Thinking Society, Michael Marshall (@MrMMarsh and @GoodThinkingSoc). There are numerous other Twitter accounts exposing quackery and campaigning against health pseudoscience, including Quackery Detector (@QuackDetector), Blue Wode (@Blue_Wode), Bob Blaskiewicz (@rjblaskiewicz), ScienceBasedMedicine (@ScienceBasedMed), Health Watch (@

HealthWatch123), medical scientist and author Paul Offit (@DrPaulOffit), medical scientist, author and campaigner against bad science, and Ben Goldacre (@bengoldacre), to mention just a few.

The darker side: the dangerous infection spread by social media

Social media, including Twitter, spread information, whether accurate or fake/false, even faster than a virus can spread an infection (Del Vicario et al., 2016). 'Online environments do exhibit polarization characteristics where misinformation can spread virally', as Caulfield et al. (2019: 53) point out; furthermore, 'health and science information on these platforms is often problematic'. Misinformation and harmful health messages on social media are common in contentious fields such as anti-vaccination propagation (Dunn et al., 2015; Tomeny et al., 2017), the Ebola virus outbreak (Oyeyemi et al., 2014), Lyme disease (Basch et al., 2017), and the Zika virus (Sharma et al., 2017). Studies have also shown that falsehoods can diffuse 'farther, faster, deeper and more broadly' than the truth (Oyeyemi, Gabarron & Wynn, 2014). Caulfield et al. (2019: 53) surmise 'while notions of the "echo chamber" might be overstated' (Dubois & Blank, 2018; Flaxman et al., 2013), the viral effect of social media messages is irrefutable.

This analogy of the viral effect of social media to distribute misinformation and quackery over a wide area and to potentially millions of users will be used as illustration in two case studies: news about the Ebola virus, and the anti-vaccination campaign.

Spreading false information about the Ebola virus

Vosoughi et al. (2018) have found that the spread of false health news is often much more effective than the truth. In an analysis of the media coverage on social media of the Ebola viral outbreak in 2014, Oyeyemi et al. (2014) pointed out the dangers of a combination of Ebola, Twitter and misinformation. Similarly, Van Rooyen (2017: 11) found 'seemingly equal amounts of media coverage were devoted to the positive role [...] that social media were playing in aiding the fight against the pandemic and the negative role [...] that social media were playing by allowing for the rapid, rampant spread of misinformation about the disease'.

Blair (2014) emphasises the dichotomy in social media platforms as these forms of media messages are

> as unregulated as they are democratizing. The Ebola outbreak has unveiled a darker side of social media – the voracious spread of misinformation. Rumored preventatives and cures rapidly gain traction online as desperate West Africans search for any method to counteract the thus-far untreatable disease. Eating raw onion, eating koala-nut, or drinking coffee have all surfaced as solutions. In Nigeria, two people died from drinking salt water – making misinformation in that country half as deadly as the disease itself. The rumored cure has hospitalised dozens more. The ill-informed noise on social media has made it difficult for legitimate sources, such as the Centers for Disease Control (CDC) or the World Health Organization (WHO) to make their voices heard.

Risen (2014) argues that the 'internet has shown a dark side during the Ebola outbreak as well by making it easier for people to spread misinformation about the disease. Online scam artists are selling products they claim can prevent or cure Ebola using everything from silver, herbal remedies or even snake venom'.

Tessler (2014: n.p.), a consumer education specialist at the US Federal Trade Commission (FTC), emphasises that at the FTC they have learnt that 'scams often follow the news – especially when there's a health scare in the headlines [...] Banking on fear, scam artists are making unsubstantiated claims that products containing everything from silver to herbal oils and snake venom can cure or prevent Ebola'.

Referring to the Ebola outbreak in Africa in 2014, Van Rooyen (2017: 10) cites the *Time* magazine journalist Victor Luckerson

(2014a) who wrote (although specifically only referring to the American media's perception and interpretation of the epidemic after the death of Thomas Eric Duncan, the first person in the US to die of Ebola shortly after travelling from West Africa):

> Based on Facebook and Twitter chatter, it can seem like Ebola is everywhere. Following the first diagnosis of an Ebola case in the United States on Sept. 30, mentions of the virus on Twitter leapt from about 100 per minute to more than 6,000 [...] Trying to stem the spread of bad information online actually shares many similarities with containing a real-world virus. Infected Internet users, who may have picked up bogus info from an inaccurate media report, another person on social media or word-of-mouth, proceed to 'infect' others with each false tweet or Facebook post.

In another article, Luckerson (2014b) continues: 'Between 16 September and 6 October 10.5 million tweets mentioning the word "Ebola" were recorded' (cited by Van Rooyen, 2017: 11).

Yet Twitter can also be a counter to viral misinformation, as pointed out by Murdock (2014) with regard to the handling of the disease in Nigeria:

> The Nigerian government says communication is its first line of defense against Ebola. With no known cure and new fears about a potentially infected corpse found at a mortuary, health officials are Facebooking, Tweeting and writing radio jingles in an effort to reach everyone in Africa's most populous country. Their main message is 'Wash your hands'. [...] Health officials are also posting information about how the disease spreads and numbers to call for questions or to report illness on their Facebook page, that are being Tweeted by other agencies, like the Nigerian Police.

Summarising, Van Rooyen (2017) argues the 'Ebola outbreak clearly showed that Twitter can play a massive role in the dissemination of science-related news and information, both accurate and inaccurate, especially if people believe the information can impact their lives [...] It also serves to demonstrate Twitter's enormous potential for good science communication through the viral tweeting and retweeting of sound science'.

The anti-vaccination viral infection on Twitter

The vast influence of social media in the viral spread of health quackery is nowhere more visible than in the Andrew Wakefield MMR-autism debacle, following the publication of a small study in *The Lancet* (1998: 637–641, retracted February 2010).

The final episode in the saga is the revelation that Wakefield et al. were guilty of deliberate fraud (they picked and chose data that suited their case; they falsified facts). The *British Medical Journal* has published a series of editorials on the exposure of the fraud (Deer, 2006, 2011; *British Medical Journal*, 2011; Couzin-Frankel, 2011), which appears to have taken place for financial gain. As Sathyanarayana Rao and Andrade (2011: 95) point out, 'It is a matter of concern that the exposé was a result of journalistic investigation, rather than academic vigilance followed by the institution of corrective measures. Readers may be interested to learn that the journalist on the Wakefield case, Brian Deer (Deer, 2006, 2011), had earlier reported on the false implication of thiomersal (in vaccines) in the etiology of autism. However, Deer had not played an investigative role in that report'.

Despite Wakefield's fall from grace in the scientific community (he may not practice as a medical doctor in the UK), more than two decades later, through the viral spread of his anti-vaccination messages on social media and the support he gets from ill-informed celebrities, outbreaks of measles and other vaccination preventable diseases have led Twitter to 'announce that it would be launching a new tool in search that would prompt users to head to vaccines.org, which is run by officials' at the US Department of Health and Human Services. 'Over the past few months, social media companies like Facebook and Twitter have faced intense pressure from lawmakers and the public to remove anti-vaccination propaganda from their platforms' (Kelly, 2019).

Wakefield's influence to continue spreading false and dangerous information, despite widely being discredited by various scientific studies of which the most recent comprehensive study in Denmark (Hansen et al., 2019) about vaccinations, is substantially enabled and strengthened by the role celebrities play. Jenny McCarthy, Jim Carrey, Robert de Niro, Jessica Biel, Oprah Winfrey, 'one of the most powerful enablers of cranks on the planet' (Belluz, 2018) and former US Congressman Robert Kennedy are examples of these influential voices who spread dangerous misinformation about vaccinations (Claassen, 2019b).

The anti-vaccination campaign has a serious effect on the spread of preventable diseases, as reported by McNeil (2019: n.p.): 'Measles continues to spread in the United States, federal health officials said on Monday, surpassing 700 cases this year as health officials around the country sought aggressive action to stem the worst outbreak in decades.'

The resurgence of polio in northern Nigeria under the influence of Muslim religious leaders (Kapp, 2003), the serious outbreak of measles during May 2019 in New York City (McKinley, 2019; McNeil, 2019) also because of religious traditions in an ultra-orthodox section of the Jewish community, and numerous other examples of the devastation the anti-vaccination movement is causing in the US, Australia, the UK, Germany, India, Africa and elsewhere (Infectious Disease Advisor, 2018), have become 'a litmus test not only for quality science journalism but journalism in general' (Claassen, 2019b: n.p.). As the Infectious Disease Advisor (2018: n.p.) emphasises, the 'anti-vaccine movement has proliferated over recent years, in part because of its most vocal proponents using social media to churn out often misleading information'.

The media's exposure to the pseudoscientific and quackery views of celebrities has aggravated the role social media and specifically Twitter play in virally spreading misleading health information. Julie Gunlock, a senior fellow at the Independent Women's Forum and leader of the organisation's *Culture of Alarmism Project,* emphasises in an opinion piece in the *Wall*

Street Journal that the 'anti-vaccine hysteria Ms Winfrey helped incubate was more dangerous than mere "fake news". It actually put people's lives at risk' (Gunlock, 2018).

In summary, anti-vaccination messages and their acceptance have virally spread through the public sphere, a prime example of Habermas's theory (1991). As Jolley and Douglas (2014) and Stein (2017) point out, the anti-vaccination movement has been characterised by conspiracy theories abounding in the movement, as elaborated by the warning of the Infectious Disease Advisor (2018: n.p.):

> Anti-vaccination rhetoric has become part of the mainstream dialogue regarding childhood vaccination, and social media is often employed to foster online spaces that strengthen and popularise anti-vaccination theories [...] Conspiracy theories have become endemic among anti-vaccination groups. These sentiments have been compounded in recent years by decreased trust in the institutions that manufacture or distribute vaccines. The effect of vaccination refusal on public health is particularly challenging when misinformation is disseminated through social media. Thought influencers in the anti-vaccine movement include doctors, celebrities, community organisers, and 'mommy bloggers' who collectively speak to an audience of about 7 million Facebook followers. The potential for disseminating harmful health-related information through social media seems to be at an all-time high.

Conclusion

For journalists, often the most accessible and visible direct science communicators to the general public, the challenge is to make a clear distinction between, on the one hand, claims about health issues and cures by pseudoscientists and often fraudulent marketers, not based on evidence and without undergoing rigorous clinical trials and peer review, and, on the other hand, valid scientific research and findings based on clinical trials which are properly peer-reviewed. To make sense of all the health claims flooding Twitter and other social media platforms, informed journalists and scientists have become vital conduits to inform the public of the validity of claims, to point out the harm quackery and pseudoscientific assertions can cause. That means that there can be immense benefit for society in general if scientists and science journalists have a presence on Twitter and expose the fallacies of quacks and the harm they can cause.

The role the media play has often been to provide platforms to spread pseudoscience and this phenomenon does not only damage the media's reputation in the eyes of scientists (Claassen, 2011: 361), but also proliferates on Twitter and Facebook where untested and unverified health information is virally spread. From the perceptive of scientists, 'There has thus arisen a view of the media as a "dirty mirror" held up to science, an opaque lens unable adequately to reflect and filter scientific facts' (Bucchi, 2004: 109). As Park (2000: 67) aptly describes, the marketing of pseudoscience and quackery as valid science, reinforces 'a sort of upside-down view of how the world works, leaving people vulnerable to predatory quacks. It's like trying to find your way around San Francisco with a map of New York'.

This situation is aggravated by the serious misconception of journalists that the view of scientists and non-scientists should be balanced. 'The he said/she said framework of modern journalism ignores (the) reality' that '(u)ntil a claim passes that judgement – that *peer review* – it is only that, just a claim' (Oreskes & Conway, 2011: 269). This misinterpretation of the question of balanced reporting regarding science and pseudoscience becomes a dire problem when false health claims are virally spread on Twitter and other social media platforms. Rensberger (2002: n.p.) emphasises the link between the need for evidence, its trustworthiness and the weight of that evidence:

> Science demands evidence, and some forms of evidence are worth more than others are. A scientist's authority should command

> attention but, in the absence of evidence, not belief [. . .] Balanced coverage of science does not mean giving equal weight to both sides of an argument. It means *apportioning weight according to the balance of evidence* (author's emphasis).

In dealing with 280-character messages on Twitter, science reporters have to adopt the mode of evaluating evidence that is so integrally part of scientists' equipment; it is a moral obligation to weigh all claims on a scale of evidence, to test the veracity of marketing messages by pseudoscientists. This is one of the most serious challenges for 21st-century journalists but also scientists in the age of social media, that the viral spread of misinformation by harmful quacks be countered timeously and without hesitation.

Furthermore, the media should be vigilant about the dangers of celebrity capture when it comes to science and the role celebrities play on Twitter. Science is too intricate to be left to ignoramuses, too often scientifically illiterate journalists hold up a 'dirty mirror' (Bucchi, 2004: 109) of science, reflecting through an opaque lens celebrities' view of life, 'a tale told by an idiot, full of sound and fury, signifying nothing', as Oreskes and Conway (2011: 274) quote *Macbeth* (Act 5, Scene 5) so applicably.

References

Allgaier, J., Dunwoody, S., Brossard, D., Lo, Y. & Peters, H.P. (2013). Journalism and social media as means of observing the contexts of science. *BioScience*, 63(4), 284–287. https://doi.org/10.1525/bio.2013.63.4.8.

Anderson, A. A., Brossard, D. & Scheufele, D. A. (2010). The changing information environment for nanotechnology: Online audiences and content. *Journal of Nanoparticle Research*, 12(4), 1083–1094. https://www.ncbi.nlm.nih.gov/pmc/articles/PMC2988218/. DOI: https://doi.org/10.1007/s11051-010-9860-2.

Basch, C.H., Mullican, L.A., Boone, K.D., Yin, J., Berdnik, A., Eremeeva, M.E. & Fung, I.C. (2017). Lyme disease and YouTube™: A cross-sectional study of video contents. *Osong Public Health and Research Perspectives*, 8(4), 289–292.

Bauer, M. (2000). Science in the media as a cultural indicator: Contextualising surveys with media analysis. In M. Dierkes & C. von Groete (eds), *Between Understanding and Trust: The public, science and technology* (pp. 157–178). Amsterdam: Harwood Academic.

Bauer, M. (2008). Survey research and the public understanding of science. In M. Bucchi & B. Trench (eds), *Handbook of Public Communication of Science and Technology*, (pp. 111–129). London: Routledge.

Bell, E. (2016). Facebook is eating the world. *Columbia Journalism Review*, 7 March. http://www.cjr.org/analysis/facebook_and_media.php. Accessed 15 June 2018.

Belluz, J. (2018). Oprah's long history with junk science. She may be the most powerful crank enabler on the planet, 9 January; https://www.vox.com/science-and-health/2018/1/9/16868216/oprah-winfrey-pseudoscience.

Bik, H. M. & Goldstein, M. C. (2013). An introduction to social media for scientists. *PLOS Biology*, 11(4). http://journals.plos.org/plosbiology/article?id=10.1371/journal.pbio.1001535.

Birch, H. (2011). The social web in science communication. In D. J. Bennett & R. C. Jennings (eds), *Successful Science Communication: Telling it like it is* (pp. 280–293). Cambridge, UK: Cambridge University Press. https://doi.org/10.1017/CBO9780511760228.024.

Blair, E. (2014). #Ebola lessons: How social media gets infected. *Information Week*. http://www.informationweek.com/software/social/-ebola-lessons-how-social-media-gets-infected/a/d-id/1307061.

Bombaci, S. P., Farr, C. M., Gallo, H. T., Mangan, A. M., Stinson, L. T., Kaushik, M. & Pejchar, L. (2016). Using Twitter to communicate conservation science from a professional conference. *Conservation Biology*, 30(1), 216–225. https://doi.org/10.1111/cobi.12570.

Boseley, S. (2009, 14 October). Science writer Simon Singh wins ruling in chiropractic libel battle. *The Guardian*. https://www.theguardian.com/media/2009/oct/14/simon-singh-chiropractors-appeal.

British Royal Society (BRS) (1995). *The Public Understanding of Science*. London: BRS.

British Royal Society (2007). *International Indicators of Science and the Public*. London: BRS.

Bronner, G. (2011). *The Future of Collective Beliefs*. Oxford: The Bardwell Press.

Bucchi, M. (1998). *Science and the Media: Alternative routes in scientific communication*. London: Routledge.

Bucchi, M. (2004). *Science in Society: An introduction to social studies of science*. London: Routledge.

Caulfield, T., Marcon, A. R., Murdoch, B., Brown, J. M., Perrault, S. T., Jarry, J., et al. (2019). Health misinformation and the power of narrative messaging in the public sphere. *Canadian Journal of Bioethics Revue Canadienne de bioéthique*, 2(2). https://www.erudit.org/en/journals/bioethics/2019-v2-n2-bioethics04449/1060911ar.pdf.

Claassen, G. (2011). Science and the media in South Africa: Reflecting a 'dirty mirror'. *Communicatio*, 37(3), 351–366. http://dx.doi.org/10.1080/02500167.2011.622288.

Claassen, G. (2019a). Science, morality and the media: Complicity in spreading pseudoscience, or watchdog of the public? In C. Jones & J. van den Heever (eds), *Moral Issues in the Natural Sciences and Technology* (pp. 199–218). Cape Town: AOSIS.

Claassen, G. (2019b, June 25). Science denialism is a litmus test for quality journalism. *News24*. https://www.news24.com/Columnists/GeorgeClaassen/science-denialism-is-a-litmus-test-for-quality-journalism-20190625.

Colditz, J. B., Chu, K. H., Emery, S. L., Larkin, C. R., James, A. E., Welling, J. & Primack, B. A. (2018). Toward real-time infoveillance of Twitter health messages. *American Journal of Public Health*, 108(8), 1009–1014. https://www.ncbi.nlm.nih.gov/pubmed/29927648.

Collins, K., Shiffman, D. & Rock, J. (2016). How are scientists using social media in the workplace? *PLoS One*. https://doi.org/10.1371/journal.pone.0162680.

Couzin-Frankel, J. (2011, 6 January). *British Medical Journal* charges fraud in autism-vaccine paper, *Science*. https://www.sciencemag.org/news/2011/01/british-medical-journal-charges-fraud-autism-vaccine-paper.

Deer, B. (2006, 31 December). MMR doctor given legal aid thousands. *The Sunday Times*. http://www.timesonline.co.uk/tol/news/uk/article1265373.ece.

Deer, B. (2011). How the vaccine crisis was meant to make money. *British Medical Journal*, 342, 136–142.

Del Vicario, M., Bessi, A., Zollo, F., Petroni, F., Scala, A., Caldarelli, G., et al. (2016). The spreading of misinformation online. *Proceedings of the National Academy of Sciences*, 113(3), 554–559.

Dentzer, S. (2009). Communicating medical news: Pitfalls of health care journalism. *New England Journal of Medicine*, 360(1), 1–3.

Dubois, E. & Blank, G. (2018). The echo chamber is overstated: The moderating effect of political interest and diverse media. *Communication & Society*, 21(5), 729–745.

Dunn, A. G., Leask, J., Zhou, X., Mandl, K. D. & Coiera, E. (2015). Associations between exposure to and expression of negative opinions about human papillomavirus vaccines on social media: an observational study. *Journal of Medical Internet Research*, 17(6). https://www.ncbi.nlm.nih.gov/pmc/articles/PMC4526932/.

Ernst, E. & Smith, K. (2018). *More Harm than Good? The moral maze of complementary and alternative medicine.* Munich: Springer.

Flaxman, S., Goel, S. & Rao, J. M. (2013). Ideological segregation and the effects of social media on news consumption. Microsoft Research. https://pdfs.semanticscholar.org/768e/b9576a9a478c95e8ed3434ea4752c4098aa7.pdf?_ga=2.237553280.1605003808.1566915035-456935960.1565264758.

Franzen, M., Rödder, S. & Weingart, P. (2007). Fraud: Causes and culprits as perceived by science and the media. *Embo Reports*, 8, 3–7. https://www.embopress.org/cgi/doi/10.1038/sj.embor.7400884.

Gastrow, M. (2015). Science and the social media in an African context: The case of the Square Kilometre Array telescope. *Science Communication*, 37(6), 703–722. https://doi.org/10.1177/1075547015608637.

Gray, J. (2019, 14 August). Why liberals now believe in conspiracies. *New Statesman*. https://www.newstatesman.com/politics/uk/2019/08/why-liberals-now-believe-conspiracies.

Gunlock, J. (2018, 10 January). Oprah's 'truth' and its potentially deadly consequences. *Wall Street Journal.* https://www.wsj.com/articles/oprahs-truth-and-its-potentially-deadly-consequences-1515628059.

Habermas, J. (1991). *The Structural Transformation of the Public Sphere: An inquiry into a category of bourgeois society*. Boston, MA: MIT Press.

Hall, N. (2014). The Kardashian Index: A measure of discrepant social media profile for scientists. *Genome Biology,* 15, 424, https://genomebiology.biomedcentral.com/articles/10.1186/s13059-014-0424-0.

Hansen, N. D., Mølbak, K., Cox, I. J. & Lioma, C. (2019). Relationship between media coverage and measles-mumps-rubella (MMR) vaccination uptake in Denmark: Retrospective study. *JMIR Public Health and Surveillance,* Jan-March. https://www.ncbi.nlm.nih.gov/pmc/articles/PMC6364207/.

Hartz, J. & Chappell, R. (1997). *Worlds Apart: How the distance between science and journalism threatens America's future.* Nashville, Tennessee: First Amendment Centre.

Hesmondhalgh, D. (2008). *The Media and Social Theory*. London: Routledge.

Infectious Disease Advisor (2018, 31 October). Social medicine: The effect of social media on the anti-vaccine movement. https://www.infectiousdiseaseadvisor.com/home/topics/prevention/social-medicine-the-effect-of-social-media-on-the-anti-vaccine-movement/.

Jarreau, P. B. (2016). Using Twitter to interact, but science communication to preach, *SciLogs*. http://www.scilogs.com/from_the_lab_bench/using-twitter-to-interact-but-science-communication-to-preach/.

Jarvis, W. T. (1997, 29 May). How quackery harms cancer patients. *Quackwatch*. https://www.quackwatch.org/01QuackeryRelatedTopics/harmquack.html.

Jolley, D. & Douglas, K. M. (2014). The effects of anti-vaccine conspiracy theories on vaccination intentions. *PloS One*. https://www.ncbi.nlm.nih.gov/pmc/articles/PMC3930676/.

Joubert, M. & Guenther, L. (2017). In the footsteps of Einstein, Sagan and Barnard: Identifying South Africa's most visible scientists. *South African Journal of Science,* 113(11/12). https://doi.org/10.17159/sajs.2017/20170033.

Joubert, M. (2018). Factors Influencing the Public Communication Behaviour of Publicly Visible Scientists in South Africa. Unpublished doctoral thesis, Stellenbosch University. http://scholar.sun.ac.za/handle/10019.1/103268.

Joubert, M. (2019). New policy commits South Africa's scientists to public engagement. Are they ready? *The Conversation Africa*. https://theconversation.com/new-policy-commits-south-africas-scientists-to-public-engagement-are-they-ready-114832

Joubert, M. & Costas, R. (2019). Getting to know science tweeters: A pilot analysis of South African Twitter users tweeting about research articles. *Journal of Altmetrics.* https://www.journalofaltmetrics.org/article/10.29024/joa.8/.

Kapp, C. (2003). Surge in polio spreads alarm in northern Nigeria. *The Lancet,* 1631.

Kelly, M. (2019, 14 May). Twitter fights vaccine misinformation with new search tool. *The Verge*. https://www.theverge.com/2019/5/14/18623494/twitter-vaccine-misinformation-anti-vax-search-tool-instagram-facebook.

Laudan, L. (1983). The demise of the demarcation problem. In R. S. Cohen & L. Laudan (eds), *Physics, Philosophy and Psychoanalysis: Essays in honor of Adolf*

Grünbaum (pp.111–127). Boston Studies in the Philosophy of Science. Dordrecht: Reidel.

Luckerson, V. (2014a). Fear, misinformation, and social media complicate Ebola fight. *Time.* http://time.com/3479254/ebola-social-media/.

Luckerson, V. (2014b). Watch how word of Ebola exploded in America. *Time.* http://time.com/3478452/ebola-twitter/.

McKinley, J. (2019, 13 June). Measles outbreak: NY eliminates religious exemptions for vaccinations, *The New York Times.* https://www.nytimes.com/2019/06/13/nyregion/measles-vaccinations-new-york.html.

McNeil, D. G. (2019, April 29). Measles cases surpass 700 as outbreak continues unabated, *The New York Times.*

Mooney, C. & Kirshenbaum, S. (2009). *Unscientific America: How scientific illiteracy threatens our future.* New York: Basic Books.

Mudde, S. E. (2019). A Tale of Two Citizens: How South Africa's most visible scientists use Twitter to communicate with the public. Unpublished masters thesis, Stellenbosch University.

Murdock, H. (2014). Nigeria using Facebook, Twitter to inform people about Ebola. *Voice of America.* http://www.voanews.com/content/nigeria-embarks-on-mass-communications-to-prevent-ebola-spread/1969796.html.

Nagy, P., Wylie, R., Eschrich, J. & Finn, E. (2018). The enduring influence of a dangerous narrative: How scientists can mitigate the Frankenstein myth. *Journal of Bioethical Inquiry,* 15(2), 279–292. https://www.ncbi.nlm.nih.gov/pubmed/29525895.

Novak, R. (2015). Communicating #science on Twitter works. Colorado State University. http://source.colostate.edu/communicating-science-on-twitter-works/.

Nelkin, D. (1995). *Selling Science: How the press covers science and technology.* New York: W.H. Freeman.

Oreskes, N. & Conway, E. M. (2011). *Merchants of Doubt: How a handful of scientists obscured the truth on issues from tobacco smoke to global warming.* London: Bloomsbury.

Oyeyemi, S. O., Gabarron, E. & Wynn, R. (2014). Ebola, Twitter, and misinformation: A dangerous combination? *British Medical Journal,* 349. https://www.bmj.com/content/349/bmj.g6178.

Park, R. (2000). *Voodoo Science: The road from foolishness to fraud.* Oxford: Oxford University Press.

Pigliucci, M. (2010). *Nonsense on Stilts: How to tell science from bunk.* Chicago, IL: University of Chicago Press.

Pigliucci, M. & Boudry, M. (2013). *Philosophy of Pseudoscience: Reconsidering the demarcation problem,* Chicago, IL: University of Chicago Press.

Pulido, C. M., Redondo-Sama, G., Sordé-Martí, T. & Flecha, R. (2018). Social impact in social media: A new method to evaluate the social impact of research. *PLoS One,* 13(8). https://www.ncbi.nlm.nih.gov/pubmed/30157262.

Rensberger, B. (2002). What every journalist should know about science and science reporting. *Nieman Reports.* https://niemanreports.org/articles/what-every-journalist-should-know-about-science-and-science-journalism.

Risen, T. (2014). Mobile phones, social media aiding Ebola fight. *US News*. http://www.usnews.com/news/articles/2014/10/10/phones-social-media-aiding-in-ebola-fight.

Ruse, M. (2013). *The Gaia Hypothesis: Science on a pagan planet*. Chicago, IL: University of Chicago Press.

Sathyanarayana Rao, T. S. & Andrade, C. (2011). The MMR vaccine and autism: Sensation, refutation, retraction, and fraud. *Indian Journal of Psychiatry*, 53(2), 95–96. https://www.ncbi.nlm.nih.gov/pmc/articles/PMC3136032/.

Shakespeare, William. (1985). *The Complete Works of William Shakespeare*. New York: Nelson Doubleday.

Sharma, M., Yadav, K., Yadav, N. & Ferdinand, K. C. (2017). Zika virus pandemic: Analysis of Facebook as a social media health information platform. *American Journal of Infection Control*, 45(3), 301–302.

Shukla, R. & Bauer, M. (2007). *The Science Culture Index (SCI): Construction and validation. A comparative analysis of engagement, knowledge and attitudes to science across India and Europe*. London: The British Royal Society.

Singh, S. & Ernst, E. (2008). *Trick or Treatment: Alternative medicine on trial*. London: Corgi.

Stein, R.A. (2017, 5 December). The golden age of anti-vaccine conspiracies. *Germs*. https://www.ncbi.nlm.nih.gov/pmc/articles/PMC5734925/.

Steinman, H. (2014, 19 June). Medicine rules are insulting to consumers. *CamCheck*. https://www.camcheck.co.za/medicines-rules-are-insulting-to-consumers/.

Steinman, H. (2018, 16 July). This natural trick can cure your cancer. *CamCheck*. https://www.camcheck.co.za/this-natural-trick-can-cure-your-cancer/.

Su, L. Y. F., Scheufele, D. A., Bell, L., Brossard, D. & Xenos, M. A. (2017). Information-sharing and community-building: Exploring the use of Twitter in science public relations. *Science Communication*, 39(5), 569–597.

Tessler, C. (2014, 9 October). Scammers bank on Ebola fears. Federal Trade Commission. https://www.consumer.ftc.gov/blog/2014/10/scammers-bank-ebola-fears.

The Lancet (2010). Retraction – Ileal-lymphoid-nodular hyperplasia, non-specific colitis, and pervasive developmental disorder in children. 375, 445. https://www.thelancet.com/journals/lancet/article/PIIS0140-6736%2810%2960175-4/fulltext.

Tomeny, T. S., Vargo, C. J. & El-Toukhy, S. (2017). Geographic and demographic correlates of autism-related anti-vaccine beliefs on Twitter, 2009–15. *Social Science & Medicine*, 191, 168–175.

Van Rooyen, R. S. (2017). *Science Communication and the Nature of the Social Media Audience: Breaking and spreading of science news on Twitter in the South African context*. Unpublished master's thesis, Stellenbosch University. http://scholar.sun.ac.za/handle/10019.1/100908.

Vosoughi, S., Roy, D. & Aral, S. (2018). The spread of true and false news online. *Science*, 359(6380), 1146–1151.

Vraga, E. K. & Bode, L. (2017). Using expert sources to correct health misinformation in social media. *Science Communication*, 39(5). https://www.researchgate.

net/publication/319853685_Using_Expert_Sources_to_Correct_Health_Misinformation_in_Social_Media.
Webster, F. (2006). *Theories of the Information Society*. London: Routledge.
West, A. (2019, 13 May). A 23-Year record of vigilance against questionable healthcare claims and their financial implications. *Quackwatch*. https://www.cardrates.com/news/quackwatch-protects-against-questionable-health-care-claims-and-their-financial-implications/.
Yeo, S. K., Cassiatore, M. A., Brossard, D., Scheufele, D. A. & Xenos, M. A. (2014). Twitter as the social media of choice for sharing science. 13th International Public Communication of Science and Technology Conference, Salvador, Brazil. https://www.researchgate.net/publication/259891600_Twitter_as_the_social_media_of_choice_for_sharing_science.

8 The amplification of uncertainty: The use of science in the social media by the anti-vaccination movement

François van Schalkwyk

Two relatively recent developments are, in one way or another, changing the science communication environment. The first is the progression towards a more accessible science (Friesike et al., 2015; Leonelli et al., 2015) while the second is the pervasiveness of the social media in our daily lives (Schäfer, 2017; Southwell, 2017; Williams, 2018). Both take place in a broader social context of persistently high levels of distrust in public institutions (Edelman Trust Barometer, 2019; Ortiz-Ospina & Roser, 2016; Winowatan et al., 2019) and new networked social configurations (Castells, 1996, 2009). Some suggest that we are witnessing, in some cases at least, the pollution of the science communication environment (Kahan, 2016). Others express concern over the strategic use of science in the social media for political and economic ends (Weingart, 2017).

The chapter begins with a discussion on communication networks, trust, open science and the norms of science to frame its overarching line of enquiry, i.e. the observable effects of the intersection between science and the social media as they relate to the communication of science. The case of the anti-vaccination movement is put forward as appropriate to explore this intersection because the movement is attentive to science (Bean, 2011; Bennato, 2017; Kata, 2012; Moran et al., 2016) and because its use of scientific information in its online communications presents very real health risks to society (WHO, 2019).

Communication networks

In 2018, those on social media networks numbered 2.23 billion active users on Facebook[1] (Statsita, 2018) and 335 million on Twitter[2] (Statista, 2018). Digital media and infrastructure create an integrated, networked environment based on flows of information. Increasingly, this environment provides the primary setting for human agency (Castells, 1996, 2009).

According to Castells, the basic elements of the network society are not material, but the intangible flows of information produced by and processed through media: Information to communicate among people, to control processes, to check and re-evaluate existing information, and to produce more and new information (Stalder, 2006).

It is not that networks are new but that digital information networks introduce new realities of communication and therefore, by implication, of social relations. The space of flows brings distant elements (things and people) into an interrelationship that is characterised by being continuous and in real time (Castells, 1996). From a historical perspective, this conflation of spatial and temporal separation is new.

According to Castells (2009), there are multiple global communication networks, the contours of which are not always sharply defined. Networks overlap and are influenced by one another, and networks compete and defend themselves. One cannot therefore understand one network without reference to other networks, although Castells argues that it is the global financial network that dominates in the current global capitalist economic dispensation (Castells, 2009).

---

1 As at the second quarter of 2018.

2 Ibid.

A network is defined by the program that assigns the network its goals and its rules of performance; in other words, the core logic of the network. A network's program consists of codes for the evaluation of performance and criteria for success or failure in the network. To transform the outcomes of any specific network, a new program emanating from outside the network must displace the existing program of the network, and control over communication is a key determinant in the outcome of any attempted displacement (Castells, 2009).

For science, the emergence and entrenchment of digital communication networks in society have had a series of impacts on its communication. The digitisation of the traditional print media and the advent of online social networks have disrupted the communication of science (Brossard, 2013; Scheufele, 2013; Southwell, 2017) and are likely to continue. As socially constructed space, the relationships between social actors (and objects) in the networks of communication in the age of information is therefore key to understanding the delivery, reception, use, re-use and impact of science communication.

Trust in science

Referring specifically to science, Popper (1962) also attributes the acquisition and application of the capacity to recognise science to an immersion in a set of social processes and conventions. Any influence that impairs or impedes these social practices will degrade the ability of the public to recognise valid science and hence to fully realise its benefits. The key concepts at work are influence and validity, and both are strongly linked to trust.

How trust is established between science and its publics is poorly understood (Scheufele, 2014; Weingart & Guenther, 2016). Schäfer (2016) argues for a greater acknowledgement within the field of science communication of the role that trust plays in the intermediation process of communicating science. Weingart and Guenther (2016) argue that trust is in part a factor of intent in relation to the public good. Those whose intentions are in the public interest (for example, firemen) are trusted more than those perceived to harbour intentions that are self-promoting (for example, politicians). But the markers or social cues for establishing trust aren't always visible (Lin, 2008) or may be replaced with new cues when the communication of science is interpersonal as is the case in the social media (Southwell, 2017). It remains an open question why publics are receptive to the communications of selected non-scientific intermediaries in such networks, although some suggest a new conceptualisation of power in the form of the influence wielded by intermediaries in the network society (Muller 2017).

The media have traditionally been the primary interface between science and the public (Weingart, 2011), and it is the science journalist who has traditionally kept the public informed on the latest developments from the world of science (Schäfer, 2017). There has, however, been a decline in science journalism (Scheufele, 2013; Schäfer, 2017), an increase in the clamour for attention among a variety of would-be network programmers (Weingart & Guenther, 2016; Williams 2018), and an emergence of informal, interpersonal communication between science and its publics via social media (Southwell, 2017). Individuals and minority groups broadcast their own content, and attract and surpass the levels of attention garnered by the mass media because of the ubiquity of online communication networks such as Twitter, Facebook, YouTube and Instagram (Schäfer, 2017; Southwell, 2017) as well as the propensity of their programmers to capture our attention (Williams, 2018; Wu, 2016).

Bucchi (2018) describes this scenario as a 'crisis of mediators'. Scientific research and information are increasingly fed in real time into the public domain without being filtered by communication professionals. Unfiltered (open) science communication is directly connected to populism and social trends. As a consequence, the non-scientific public must be highly adept at discerning which communication sources of scientific information to trust (Kahan et al., 2017; Scheufele, 2013).

The verification of information flowing in communication

networks cannot always not take place; recipients take information presented to them at face value. The reason for this is a structural condition of networks – the logic or programme of the network may determine that information must flow not only constantly but rapidly, negating the possibility for fact-checking and/or deferred decision-making (Stalder, 2006). Instead of an increase in trust between actors in communication networks, trust is implicit in certain communication networks because the network demands it (Stalder, 2006).

In the case of some communication networks, trust mechanisms may be created purposefully to allow information injected into the network to be taken at face value. For example, in the global financial network, the clearing house institutionalises a system of trust designed to protect the network against external threats. Without this buffer, the exchange of information would slow down considerably because the validity of the information would have to be verified outside the network itself. The clearing house in the global financial network therefore protects the constant flow of information from being interrupted by external events which would compromise the face value of the information. Networks other than the global financial network require similar central, trusted nodes that intermediate information to ensure the functioning and the survival of the communication network.

Active, trusted nodes intermediate to ensure the functioning and the survival of the communication network by guaranteeing that information can be taken at face value (Stalder, 2006).

Open science and the norms of science

The increase in advocacy for transparency and accountability, operationalised as openness and access, stems in part from a degradation of trust in public institutions (Edelman Trust Barometer 2019; Ortiz-Ospina & Roser, 2016; Winowatan et al., 2019). This includes those institutions tasked with conducting scientific research and innovating for the development of society. The breakdown of trust in institutions has also seen the rise of new public management and the escalation of quality assurance models of organisational control (Power, 1997, 2000; Taubert & Weingart, 2017). The demands for accountability through greater transparency, oversight and measurement of public institutions are buttressed by claims of beneficial returns to society (Weingart, 2012). Open science is, from such a vantage, seen as being a necessary evolution towards improvement in the efficiency, quality and relevance of science to society (Jasanoff, 2006; Leonelli et al., 2015).

From a historical perspective, Eamon (1985) argues that there was a progressive shift from a more secretive to a more public science from the 17th century onwards, accelerated by the disruptive technology of the printing press and a concomitant reaction against hierarchical and monopolistic knowledge systems. Following, among others, the influence of science reformers such as Bacon and Hartlib; the establishment of Théophraste Renaudot's *Bureau d'adresse* in Paris in 1633 and of the Royal Society of London in 1662; and the publication of the *Philosophical Transactions* in 1665, the institutional mechanisms that would govern science as a form of 'public knowledge' were in place. According to Eamon (1985: 346), 'the ideal of public knowledge was not taken to imply then – any more than it does today – that everyone had perfectly free access to scientific knowledge. Nevertheless, the institutionalisation of science under the auspices of the Baconian programme helped to confirm the scientist's special role in society, not as the guardian of secret knowledge, but as the purveyor of new truths bearing the authority of experimental evidence. Free communication within the scientific community became the norm'.

By the mid-20th century, sociologist Robert Merton (1973) had proposed four norms guiding the social behaviour of scientists, one of which, the norm of communalism, dictates that the results and discoveries of science are not the property of the individual researcher but belong to the scientific community and to society at large. More recently, with the rise of the information age, the discourse around 'openness' has predominantly been in opposition to the extractive and restrictive positioning of knowledge as a private good (Boyle, 2003; Chan & Costa, 2005). The opposition is based on the premise that the sharing and reuse of science has become less dependent on the services offered by intermediaries such as publishers. Proponents of open science have emerged in opposition to the 'enclosure' of the products of science, or at least to their control by third parties, and advocate instead for their reuse without the impediments of cost and permissions (Evans, 2005).

While the open science movement mobilised with transformative intentions, it is not immune to commercial interests (Lawson, Gray & Mauri, 2016; Taubert & Weingart, 2017). As a result, there is a counter-movement towards utilitarian and instrumentalist 'openness', with less of a focus on the potential of openness for the advancement of science, and an increased emphasis on business models designed to mine openness and extract material value (Taubert & Weingart, 2017).

The norm of organised scepticism in science implies that all formal communication is provisional and contested, and it is common practice for majority as well as minority groups of scientists to self-organise themselves in relation to truth claims made by their peers. As in any functioning democracy, the majority tends to hold power. Choosing, temporarily at least, not to take sides, there is invariably a group of undecideds. However, when minority groups are able to leverage new communication technologies to amplify their message and garner unprecedented levels of attention in relation to their size, the likelihood of swaying the undecideds increases. In the much-publicised case of voter manipulation by Cambridge Analytica using Facebook data and aggressive and highly targeted online campaigning, this group of undecideds is described as 'the persuadables' (Amer & Noujaim, 2019).

Swaying the persuadables is less likely to play out within the scientific community because of its self-imposed system of checks and balances; a system that is self-regulated because scientists value a taken-for-granted and shared objective despite any floor crossing and factionalism: the establishment of verified truths. However, external to the scientific community, the safety net of truth-seeking falls away as publics arrange themselves into majority and minority positions around contentious social issues. The undecideds are targeted with persuasive messaging by the minorities seeking to swell their numbers; and unlike in the domain of science, the common objective of truth-seeking is replaced by ideological objectives which are agnostic to the norms of science.

New potentials in the communication of science

In politics, the potential to harvest data from social media networks, and to use those same social networks to influence the outcomes of democratic processes, has been uncovered (Amer & Noujaim, 2019; Illing, 2018; Tharoor, 2018). In the world of finance, unscrupulous investment companies target the reputations of large, listed public companies and use the network effects of online communication media to profit from short selling (Cameron, 2018). If online communication networks can be deployed to disrupt politics and finance, then it seems reasonable to ask what the potentials are for science.

However, an unquestioning faith in the potential of technology to advance society mutes the concerns expressed by socially-attuned observers. Referring to the founders of Google and Facebook as examples, Naughton (2017: n.p.) reports that 'it never seems to have occurred to them that their advertising engines could also be used to deliver precisely targeted ideological and political messages'. The founder and ex-CEO of Twitter, Even Williams, has lamented the use of the platform for unintended, confrontational and nefarious purposes by some of its users (Streitfeld, 2017).

Absent in much of the science communication literature are the potential risks of the communication of science in the online networked communication environment, although there are signs that a consideration of the risks is emerging (Bishop, 2016; Dickel & Franzen, 2016; Jasanoff, 2006; Lewandowsky & Bishop, 2016). Where the effects or impacts are considered, the emphasis is often on science itself, and on the beneficial impacts (Bishop,

2016). What should be of concern to science, as it becomes more open to its publics, are non-scientific, ideologically-motivated publics who are able to access knowledge-in-progress as part of their communication strategies aimed at destabilising established truths. Such risk may outweigh the benefits. As Jasanoff (2006: 36) writes: 'When claims have arrived at a certain degree of robustness, then asking for renewed scrutiny of the ways in which those conclusions were reached strikes many observers not as justifiable curiosity but as 'manufacturing uncertainty' for political ends. When public health and safety are at stake, such needless production of uncertainty could be not entirely frivolous but downright dangerous.'

An attentive anti-vaccination movement

An example of both the amplitude and risks made possible by online communication networks is to be found in the strategies employed by the anti-vaccination movement that has shown itself to be highly attentive to science (Moran et al., 2016).

In 2005, researchers were already aware of how the 'damage' could be escalated by online communication (Zimmerman et al., 2005). According to DiResta and Lotan (2015: n.p.), '[t]his anti-vax activity might seem like low-stakes, juvenile propaganda. But social networking has the potential to significantly impact public perception of events – and the power to influence opinions increasingly lies with those who can most widely and effectively disseminate a message. One small, vocal group can have a disproportionate impact on public sentiment and legislation.' Zimmerman et al. (2005: n.p.) state that '[w]ith the burgeoning of the internet as a health information source, an undiscerning or incompletely educated public may accept these claims and refuse vaccination of their children. As this occurs, the incidence of vaccine-preventable diseases can be expected to rise'. A legitimate concern given that 15 years later, the WHO (2019) has listed vaccine hesitancy as one of the top ten global health threats.

Scientists warn that what may seem like negligible decreases in vaccination rates can have dire health outcomes as herd immunity is compromised (Lo & Hotez, 2017). Of equal concern is that while on average vaccine rates in a country such as the US have remained stable at around 90%, the perception held by the general population is that vaccination rates are in the 70–79% range (Kahan, 2014). In countries as varied as France, Russia, Japan, Italy, Greece, Iran and Vietnam, more than 20% of the population believe vaccines to be harmful (Larson et al., 2016). These are worrying statistics given that the herd immunity threshold for most available vaccines is higher than 80%.

Changing perceptions and behaviour do not fully account for changes in vaccination rates. Constraints in the supply of vaccinations also impact vaccine coverage (Vanderslott & Roser, 2018). Nevertheless, given the evidence available, the role of communication in shaping perceptions and amplifying anti-vaccination messaging cannot be ignored; particularly if, as the US CDC suggests, 'philosophical objections' rather than supply constraints accounted for 79% of measles vaccination refusals in 2012 (CDC, 2013).

Given changes in the science communication environment and possible risks for both science and society, this chapter seeks to answer the following questions with a focus on the anti-vaccination movement: Is the anti-vaccination movement making use of scientific information in its online communications? If so, how is the movement using scientific information to promote its cause?

Methodology[3]

To determine the use of scientific information by the anti-vaccination movement in its online communications, open access journal articles on the relationship between vaccines and autism[4] were

3 See Van Schalkwyk (2019) from a more comprehensive account of the methods used in this study.

4 This specific focus on the link between vaccination and autism is supported by Moran et al.'s (2016) findings that 65.8% of 480 anti-vaccination websites in their study focused specifically on autism as a disease associated with vaccines.

identified by conducting searches of online repositories of scientific publications, by joining a known anti-vaccination Facebook group and by following an active anti-vaccination Twitter account. Limiting the selection to open access journal articles ensured that none of the articles were restricted regarding the accessibility of its content and was in keeping with the research objective of investigating the possible risk of open science.

From the sample of relevant open access journal articles, 10 were selected for closer analysis. Articles were selected in equal proportion from the online repositories (articles 1.1 to 1.5) and from the mentions of anti-vaccination accounts in the social media (articles 2.1 to 2.5). These 10 articles were selected based on their levels of online attention as indicated by each article's Altmetric Attention Score[5] (see Table 1).

Two online spheres – Twitter and the web – were analysed independently, and with some variation in the analysis owing to different affordances of each sphere, to discover whether and how those scientific articles are being used by the anti-vaccination movement.

In the case of Twitter, accounts were first categorised according to their stance (that is, whether they are anti-vaccination accounts). Thereafter, the level of activity and engagement of anti-vaccination accounts for each of the most frequently mentioned articles was determined. In the case of the web, the stance of the authors of anti-vaccination pages was already known and level of activity could not be quantified in a manner possible for the social media. Web pages were therefore only analysed for level of engagement.

The approach adopted to assess level of engagement with scientific information from open access journal articles was an attempt to go beyond views, downloads or mentions as proxies for the use of online content (Thelwall et al., 2013).

---

5 The Altmetric Attention Score is an automatically calculated, weighted count of all of the attention a research output has received across 15 different online media. For a detailed breakdown of the weightings and how the score is calculated, see https://help.altmetric.com/support/solutions/articles/6000060969-how-is-the-altmetric-score-calculated-

Table 1: Open access journal articles selected for analysis

Ref.	Title of open access journal article	Altmetric Attention Score
1.1	Imperfect vaccination can enhance the transmission of highly virulent pathogens	511*
1.2	Prevalence and characteristics of autism spectrum disorder among children aged 8 years - Autism and Developmental Disabilities Monitoring Network, 11 Sites, United States, 2012	311*
1.3	Lack of association between measles virus vaccine and autism with enteropathy	306*
1.4	GWATCH: A web platform for automated gene association discovery analysis	113*
1.5	The evolutionary consequences of blood-stage vaccination on the rodent malaria *plasmodium chabaudi*	96*
2.1	Autism occurrence by MMR vaccine status among US children with older siblings with and without autism	3,674#
2.2	Vaccines are not associated with autism: An evidence-based meta-analysis of case-control and cohort studies	2,989+
2.3	A positive association found between autism prevalence and childhood vaccination uptake across the US population	1,336+
2.4	Measles-mumps-rubella vaccination timing and autism among young African-American boys: A reanalysis of CDC data	1,048+
2.5	A two-phase study evaluating the relationship between thimerosal-containing vaccine administration and the risk for an autism spectrum disorder diagnosis in the United States	1,018+

As at 17 October 2017 * As at 18 October 2017 + As at 24 October 2017

In the subsections that follow, the methods of analysis for Twitter and web pages are described in detail.

Stance, activity and level of engagement on Twitter

Twitter data collected from the Altmetric Explorer for the 10 open access journal articles were analysed to determine (1) the number of mentions by an anti-vaccination account, (2) the number of tweets by anti-vaccination accounts, and (3) the level of engagement by anti-vaccination accounts.

The first task was to determine which of the Twitter accounts mentioning the journal articles could be classified as 'anti-vaccination'. From the Altmetric.com data, it was possible to create a list of unique Twitter accounts mentioning each of the 10 journal articles. A programmer was commissioned to develop an application[6] that could crawl the Twitter account URLs for each article. The crawler queried each Twitter account URL for a predetermined set of terms or hashtags commonly used by the anti-vaccination movement: antivax, vaxxed, vaccineinjur, vaxfax, vaccinesafety, informedconsent, vactruth. The selection of these terms was determined by: (1) their identification in previous studies investigating the use of Twitter in the anti-vaccination debate (Dredze et al., 2017; Mitra et al., 2016; Radzikowski et al., 2016); (2) additional terms noted by the researcher while creating the sample of anti-vaccination web pages referring to scientific research;[7] and (3) the frequency with which the terms are used on Twitter as indicated by Symplur Signals.[8]

The application returned the number of times each term could be found for each of the URLs. The presence of one or more hashtags was taken to indicate that a Twitter account associated with the URL is, most likely, anti-vaccination. The crawler only detects Twitter terms or hashtags that appear on the first page of a Twitter account. If an account used the hashtags in the past and those hashtags no longer appear on the first page of the account, then the URL will not return a positive result. Similarly, accounts that do not use the prescribed hashtags may nevertheless be anti-vaccination. These limitations of the crawler application mean that the crawler's results are conservative estimates of the number of likely anti-vaccination Twitter accounts.

The possibility also exists that accounts for which the crawler returns positive results may in fact be false positives because some

6 See https://dev.sbc4d.com/cdv/fsv/geturl.php

7 Milani (2016) also finds that despite the many tools available for identifying and analysing Twitter hashtags, some are still only discovered by chance in the research process. Mitra et al. (2016) point out that due to the transient nature of social media, it is not possible to rely solely on terms found to be in common use in the past.

8 See https://signals.symplur.com

pro-vaccination Twitter accounts use hashtags commonly used by the anti-vaccination movement to lure anti-vaccination accounts into an exchange (Conover et al., 2011). To account for false positive results, all positive results returned by the crawler were checked manually, and all accounts found to be pro-vaccination were recorded as such and removed from the sample.

Each account identified by the crawler was coded as either 'anti', 'pro', 'neutral' or 'unknown'. An account was deemed to be anti-vaccination if any consistent anti-vaccination sentiment was expressed in the Twitter account description, in the banner image of the account or in the most recent tweets on the first page of the account, or, failing the availability of an informative description, based on the sentiment expressed in a linked website, blog post or online document. An account was coded as neutral only if an explicit statement was found indicating impartiality and there was evidence of posts representative of both sides of the vaccination debate. Accounts were coded as unknown when it was not possible to make a determination regarding stance.

To determine the proportion of tweets attributable to anti-vaccination Twitter accounts, the anti-vaccination Twitter accounts were compared to the list of all accounts and tweets as recorded in the Altmetric.com data.

For some of the articles in the sample, Twitter mentions were found to be low. The levels of engagement analysis was therefore limited to those articles frequently mentioned by anti-vaccination accounts on Twitter, that is, articles 2.3 (812 anti-vaccination tweets); article 2.4 (672 anti-vaccination tweets) and article 2.5 (545 anti-vaccination tweets). For practical reasons, not all Tweets could be analysed for level of engagement. A simple random sample of 100 anti-vaccination tweets was generated for each of the three articles.

The determination of the level of engagement on Twitter by the anti-vaccination movement was done by reading each tweet in the Altmetric Explorer datasets for the three open access journal articles. Each Tweet was analysed using a 6-point scale of engagement. The scale was developed based on the suggestion by

Haustein et al. (2016) that those actions on the web that result in online visibility and traceability be categorised along a continuum of access, appraisal and application.

In an earlier study on the identification of Twitter audiences, Haustein and Costas (2015) also set out to measure the degree to which audiences engage with tweeted journal articles. They excluded retweets and use the dissimilarity between the content of the tweet and the title of the journal article as an indicator for engagement. They provide as reason for this approach the fact that only original content constitutes engagement and also that automated bots are frequent retweeters. The scale for level of engagement developed in this study departs from such an interpretation of retweets because although retweets indicate a low level of engagement, they nevertheless are assumed to play an important role in the online communication strategies of social movements.

Progression from access to application on the engagement continuum indicates increased levels of engagement by actors with digital objects such as web pages, images, journal articles, datasets and the like. An article may generate many tweets and retweets that mention an article, but such activity may not be the result of the content of the article. For example, a retracted article may generate many mentions to the article in relation to its retraction, but such activity is not necessarily indicative of engagement with the content of the article. The scale for level of engagement attempts to measure increasing levels of engagement in relation to the content of each article rather than in relation to the degree of activity on Twitter.

The scale was tested and refined using tweets for article 1.1 in order to produce the scale in Table 2.

Tweets that no longer existed or to which access was restricted by Twitter, were included in the sample of 100 tweets but could not be analysed for obvious reasons.

Table 2: Scale for level of engagement with journal articles in anti-vaccination tweets

ENGAGEMENT →					
ACCESS		APPRAISAL		APPLICATION	
1 LOW	2 LOW	3 MEDIUM	4 MEDIUM	5 HIGH	6 HIGH
Retweet OR Tweet that is copied from an earlier tweet OR broadcasting existing tweet to other accounts	Tweet article title OR tweet link and hashtags OR reply to existing tweet with link and hashtags	Tweet direct quotation from article abstract or summary	Tweet a description of the article findings in own words OR a direct quotation from the body of the article	Tweet consists of an interpretative statement or graphic pertaining to the article content	Tweet consists of an interpretative statement followed by a ciscussion thread consisting of at least a reply from another user and a response from the author of the tweet in which content from the article is used to substantiate the author's position

Level of engagement on the web

Using snowball sampling by following anti-vaccination accounts on the social media, 167 web pages were identified that made reference to a scientific source of one type or another. Of these, 70 pages included article digital object identifiers (DOIs) or PubMed IDs but only 34 web pages provided either DOIs or PubMed IDs to *full-text* open access journal articles.

The determination of level of engagement by members of the anti-vaccination movement with open access journal articles via web pages was done by developing a 6-point scale of engagement that corresponds as closely as possible to the scales used to analyse engagement on Twitter, while taking into account differences in how content is constructed and shared on social media and web pages. As with engagement on Twitter, the scale was developed based on Haustein et al.'s (2016) suggestion that engagement on the web be categorised along a continuum of access, appraisal and application, and that progression from access to application indicates increased levels of engagement by actors with digital objects. The scale was tested and refined using three randomly selected web pages. The final 6-point scale used is presented in Table 3.

Table 3: Scale for level of engagement with journal articles on anti-vaccinations web pages

ENGAGEMENT →					
ACCESS		APPRAISAL		APPLICATION	
1 LOW	2 LOW	3 MEDIUM	4 MEDIUM	5 HIGH	6 HIGH
Republication (repost) of a previously published article or blog	Includes only the title or a direct quotation from the article abstract as a reference to the article	Includes only a direct quotation from the article abstract, plus comment(s) by the author	Includes a description of the article in own words AND/OR a direct extract from the body of the article	Includes an interpretative statement/ narrative, table or graphic pertaining to the article	Includes an interpretative statement/narrative, table or graphic pertaining to the article followed (1) by a discussion thread consisting of at least a reply from another user and a response from the author of the web page in which content from the article is used to substantiate the author's position OR (2) references to and reasoned counter-arguments to pro-vaccination articles

It is important to note that the scale for level of engagement on the web does not in any way attempt to measure or assess the validity of arguments presented by the anti-vaccination movement with reference to open access scientific journal articles; the scale only seeks to measure the level of engagement with the content of those scientific articles in the construction of arguments. Only web pages written in English were analysed for level of engagement.

Findings

The findings are presented in two parts. The first part relates to mentions made specifically by the anti-vaccination movement to 10 open access journal articles. This part includes findings on the relative size and activity of the anti-vaccination movement on Twitter and addresses the question of whether the anti-vaccination movement is in fact using scientific content in its online communications. The second part presents findings on the use of open access journal articles on Twitter and web pages by applying level of engagement as a proxy for use. The findings presented in this second part of the section address the question about how the anti-vaccination movement is using openly accessible scientific content in its online communications.

Mentions of 10 open access journal articles on Twitter

Disaggregation of attentive publics on Twitter was done by determining the number of anti-vaccination Twitter accounts in each sample of all Twitter accounts that mention one of the 10 journal articles. The findings in Table 4 show that the proportion of anti-vaccination accounts to all accounts mentioning one of the 10 articles did not exceed 18%. In other words, no more than 1 in 5 mentions to an open access journal article related to the autism-vaccination debate originated from Twitter users whose stance is anti-vaccination. The findings do nevertheless confirm that the anti-vaccination movement is accessing scientific information from open access journal articles, and inserting this information into their online communications.

Table 4: Anti-vaccination Twitter accounts mentioning an open access journal article on the topic of vaccination and autism

Article	1.1	1.2	1.3	1.4	1.5	2.1	2.2	2.3	2.4	2.5
No. of unique accounts	310	123	58	81	25	3 187	2 775	1 567	931	1,397
No. of verified anti-vaccination accounts	36	1	1	0	4	35	40	218	166	157
% anti-vaccination accounts	11.6%	0.8%	1.7%	0%	16.0%	1.1%	1.4%	13.9%	17.8%	11.2%

Table 4 also shows that articles mentioned fall into two broad groups: one group of 5 articles (1.1, 1.5, 2.3, 2.4 and 2.5) in which the mentions by anti-vaccination accounts was found to be between 11% and 18% relative to all unique accounts mentioning the article on Twitter, and a second group of 5 articles (1.2, 1.3, 1.4, 2.1 and 2.2) in which fewer than 2% of mentions originated from anti-vaccination accounts.

Table 5 shows that based on textual analysis of article titles and abstracts (Van Schalkwyk, 2019), there is a relationship between the proportion of anti-vaccination accounts mentioning an article and the indicative stance of the article vis-à-vis vaccination. Unsurprisingly, those articles whose titles and findings are clearly supportive of an anti-vaccination stance are more likely to be mentioned by the anti-vaccination movement than those articles that provide no support or contradict an anti-vaccination stance.

Table 5: Vaccination stance of 10 open access journal articles and proportion of Twitter anti-vaccination accounts mentioning the article

Article ref.	Indicative stance: Title	Indicative stance: Findings	% of anti-vaccination Twitter accounts that mention the article
1.1	ANTI-VAC	ANTI-VAC	11.6
1.2	NEUTRAL	PRO-VAC	0.8
1.3	PRO-VAC	PRO-VAC	1.7
1.4	NEUTRAL	NEUTRAL	0.0
1.5	ANTI-VAC	ANTI-VAC	16.0
2.1	NEUTRAL	PRO-VAC	1.1
2.2	PRO-VAC	PRO-VAC	1.4
2.3	ANTI-VAC	ANTI-VAC	13.9
2.4	ANTI-VAC	ANTI-VAC	17.8
2.5	ANTI-VAC	ANTI-VAC	11.2

Number of tweets mentioning 10 open access articles

Further analysis of the data is possible to determine the proportion of tweets (as opposed to accounts) by the anti-vaccination movement which mention of one of the 10 open access journal articles.

Table 6 shows the proportion of tweets by the anti-vaccination movement compared to all tweets that mention one of the 10 open access journal articles. The proportion of tweets varies by article and again present in two distinct groups that correspond with the two anti-vaccination Twitter account groups.

Table 6: Anti-vaccination tweets

Article	1.1	1.2	1.3	1.4	1.5	2.1	2.2	2.3	2.4	2.5
Total no. of tweets	382	131	69	93	25	3 551	3 509	2 589	2 268	2 282
No. of anti-vaccination tweets	52	1	1	0	4	45	21	812	672	545
% of anti-vaccination tweets	13.6%	0.8%	1.5%	0%	16.0%	1.3%	0.6%	31.4%	29.6%	23.9%

Figure 1 compares the proportion of anti-vaccination Twitter accounts with the proportion of anti-vaccination tweets for each of the 10 articles. The graph shows that for those articles that appear to be of interest to the anti-vaccination movement (that is, articles 1.1, 1.5, 2.3, 2.4 and 2.5), the proportion of tweets by the anti-vaccination movement is equal to or higher than the proportion of anti-vaccination Twitter accounts for the same open access journal article. The difference is most pronounced in the cases of articles 2.3, 2.4 and 2.5.

Figure 1: % of anti-vaccination accounts compared to % of anti-vaccination tweets by article

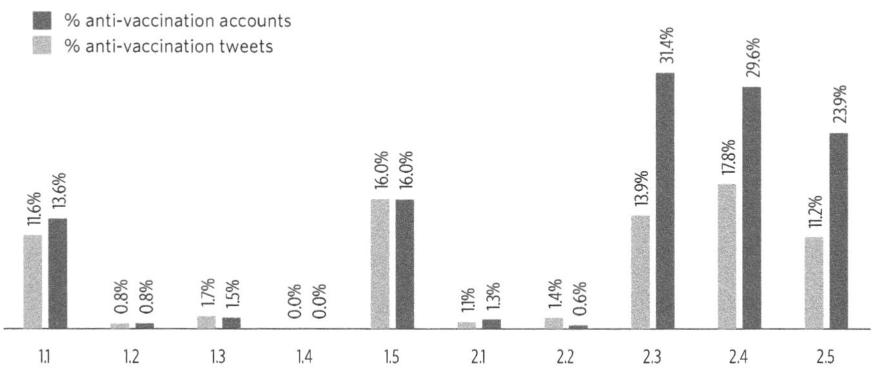

It is possible to determine which Twitter accounts are the most active in the sample of unique anti-vaccination accounts mentioning the three most frequently mentioned open access journal articles (that is, articles 2.3, 2.4 and 2.5). Figure 2 shows the frequency with which each unique account tweeted a mention to the article. There were 218 unique Twitter accounts mentioning article 2.3 in

812 tweets (Figure 2a); there were 167 unique Twitter accounts mentioning article 2.4 in 672 tweets (Figure 2b); and there were 157 unique Twitter accounts mentioning article 2.5 in 545 tweets (Figure 2c).

The data in Figure 2 show a skewed distribution in which the majority of Twitter accounts mention an article only once. The data also show that for all three articles, there are a few accounts that mention the article more than 10 times, and in all three cases there is one Twitter account that mentions the article 100 or more times.

Figure 2: Frequency of mentions by unique accounts on Twitter

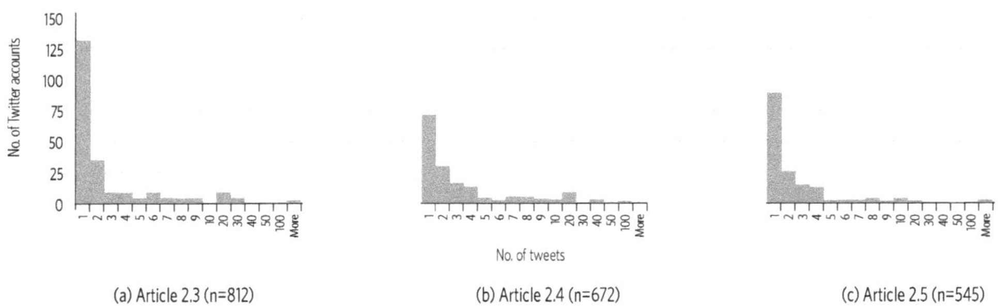

Mentions of open access journal articles on the web

During the sampling process, 75 mentions by 34 anti-vaccination web pages to full-text open access journal articles were found. This provides evidence (1) that the anti-vaccination movement is making reference to open access journal articles from its web pages, and (2) of the *potential* use of open access journal articles to support its ideology and political agenda. It is to the use of open access journal articles by the anti-vaccination movement that the next section turns its attention.

Level of engagement on Twitter

Based on the findings of the anti-vaccination movement's activity on Twitter, only three articles were selected to assess the movement's level of engagement: 2.3, 2.4 and 2.5. The selection of these three articles was determined by the fact that they are the articles that garnered the most attention from the anti-vaccination movement on Twitter.

The findings for the levels of engagement with on Twitter are shown in Figure 3. Figure 3a shows that the distribution of scores for level of engagement with Article 2.3 on Twitter fell predominantly in the access category: 90 (98%) tweets scored either 1 or 2 on the scale. Only 2 tweets fell in the appraisal category. Figure 3b shows that the distribution of scores for level of engagement with Article 2.4 on Twitter fell predominantly in the access category: 85 (98%) tweets scored either 1 or 2 on the scale. Only 2 tweets fell in the appraisal category. Figure 3c shows that the distribution of scores for level of engagement with Article 2.5 on Twitter fell predominantly in the access category: 87 (96%) tweets scored either 1 or 2 on the level of engagement scale. Only three tweets fell in the appraisal category and one tweet was found to show engagement at the level of application.

In all cases, level of engagement with the content of the three open access journal articles was found to be low. Low levels of engagement are attributable to the large proportion of retweets and reposts[9] as shown in Figure 4. In the case of article 2.3, there were 40 retweets (43%) and 31 reposts (34%). In other words, of the 92 tweets, only 21 (23%) consisted of original content. As in the case of Article 2.3, the low level of engagement with article 2.4 is explained by the finding that many of the tweets were either retweets (51, 59%) or reposts (9, 10%). Of the 87 tweets by the anti-vaccination movement, only 27 (31%) consisted of original content. The overall low level of engagement with article 2.5 is again explained by the finding that many of the tweets were either retweets (58, 64%) or reposts (19, 21%). In other words, of the 91 tweets, only 14 (15%) consisted of original content thereby limiting the possibility of higher levels of engagement.

9 A repost is defined as occurring when an account creates a new tweet or a comment that uses the exact same content as a previous tweet by the same account. A retweet occurs when an account clicks on the "retweet" affordance of an existing tweet.

Figure 3: Level of engagement on Twitter

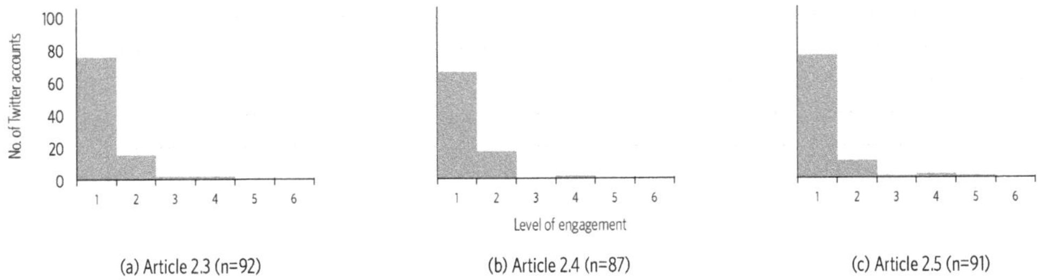

Figure 4: Tweets, retweets and reposts as an indicator of engagement

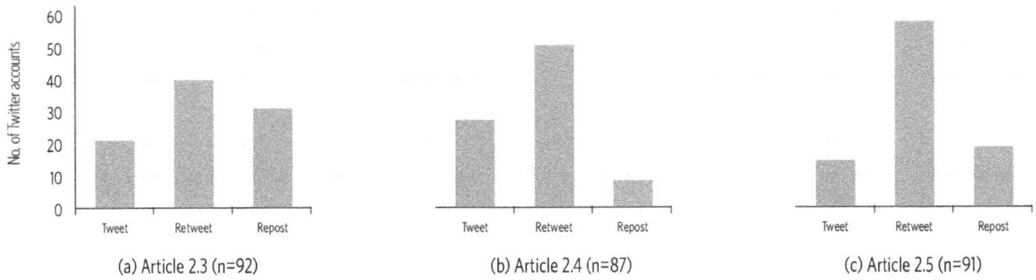

Three observations can be made in relation to level of engagement by the anti-vaccination movement with reference to Article 2.4. The first is that many of the tweets for this article do not relate to its content per se but to the fact that the article was retracted: 'An expression of concern has been published for this article. This article has been retracted. See Transl Neurodegener. 2014; 3: 22'.[10] See Figure 5 for an example of a tweet on the retraction of the article. The motivation behind these tweets is to 'prove' collusion between the CDC, the pharmaceutical industry and scientists, and there is consequently little engagement with the actual content of the article.

The second observation is the number of tweets providing an alternative link to the article post-retraction (see Figure 6). The intent of these tweets is to inform the community that the article remains accessible and, as such, available to them to support their campaign regardless of the fact that the scientific community has retracted the article from circulation. Again, the posting of a link

10 https://www.ncbi.nlm.nih.gov/pmc/articles/PMC4128611/

Figure 5: Example of a tweet regarding the retraction of article 2.4

Figure 6: Example of a tweet providing an alternative link to retracted article 2.4

does not indicate a high level of engagement with the content of the article; particularly so if the community does not engage with the scientific motivations for its retraction and elects instead to interpret the article's removal as being politically motivated.

The third observation is the use of 'broadcast' tweets in the case of mentions to Article 2.4. To illustrate: An anti-vaccination Twitter account will tag another anti-vaccination account by prefixing a (re)tweet with the Twitter handle of another member of the movement. The tweet may also include a call to action, a link or a hashtag (e.g. '#CDCwhistleblower'). Often the tagged anti-vaccination accounts will have a much larger number of Twitter followers and/or be more active on Twitter than the tweeter. In the example below (Figure 7), the tweeter had 2 888 followers while the tagged account @TannersDad had almost ten

Figure 7: Example of a tweet broadcasting article 2.4 to other Twitter accounts

times as many followers (21 400). And while the tagged accounts @ceestave and @NOWinAutism had numbers of followers similar to that of the tweeter, both accounts are highly active. The tweeter @MarcellaPiperTe had tweeted 13 200 times (since joining in June 2014), while @ceestave had tweeted 58 500 times (since May 2009) and @NOWinAustism 59 700 times (since August 2014). @TannersDad is also a highly active account with 222 000 tweets (since joining in October 2008).[11]

Article 2.5 was the only article in which a tweet was scored as being in the application category. The tweet in question was a retweet by the same account.[12] However, additional information was added to the tweet thread in the form of data published by the US Federal Drug Administration (FDA) as well as information from the journal article (including underlined text from the methods section). The first four comments comprise selected extracts the FDA on the presence of mercury in vaccines, for example, '#Flu #vaccine FLUZONE, p.18: Each 0.5 mL dose contains 25 mcg #mercury, 0.25 mL (infant) dose - 12.5 mcg mercury'. The fifth comment consists of scientific information

11 All followers and number of tweets as on 26 March 2018.

12 See the tweet in question at http://twitter.com/LotusOak/statuses/916317625407430657

extracted from the article: 'Found 7.6-FOLD Increased Risk of #AUTISM from Exposure to #Thimerosal'. It also includes an image of the results section of that article with the finding highlighted in red.[13] The tweet was deemed to indicate a level of interpretation consistent with being categorised on the 'application' end of the scale because it shows a degree of interpretation supported by content from the journal article.

Also of note is that of the tweets that mention article 2.5 in the random sample of anti-vaccination accounts, 55% (50 of 91) were retweets by the account @LotusOak. Figure 8 shows that @LotusOak tweets consistently from June 2016 to October 2017. These are not unique tweets; the majority of the tweets are retweets or reposts of the same tweet: '#STUDY: #Thimerosal-containing #Vaccines & #Autism Risk http://www.ncbi.nlm.nih.gov/pmc/articles/PMC3878266/ … It's still in multi-dose vaccines'.[14] That @LotusOak retweets the same content verbatim is further evidence of a consistently low level of engagement on Twitter as far as this journal article is concerned.

Figure 8: Tweet frequency of @LotusOak to article 2.5 (n=197)

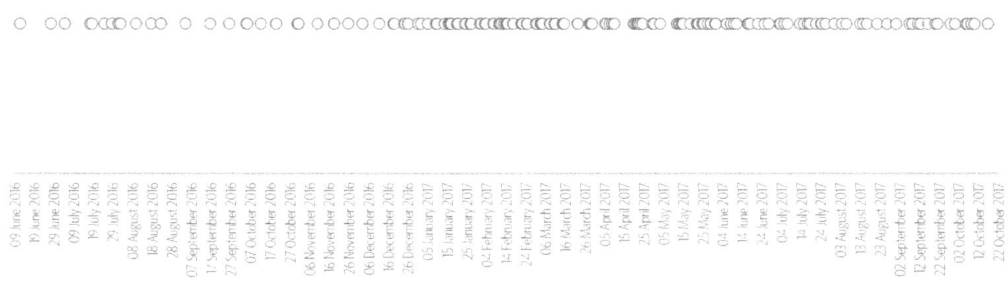

13 https://twitter.com/LotusOak/status/909814655044001794

14 https://twitter.com/LotusOak/statuses/832598920568111104

Level of engagement on the web

The findings for level of engagement by the anti-vaccination movement with open access journal articles on the topic of autism and vaccination on web pages is shown in Figure 9. The figure shows a higher level of engagement compared to the levels of engagement on Twitter: 13 (38%) web pages fall into the access category, 13 (38%) fall into the appraisal category, and 8 (24%) fall into the application category. The relatively high number of web pages in both the appraisal and the application clusters differentiates the findings on the level of engagement on the web with those on Twitter.

Figure 9: Level of engagement with open access journal articles on the web (n=34)

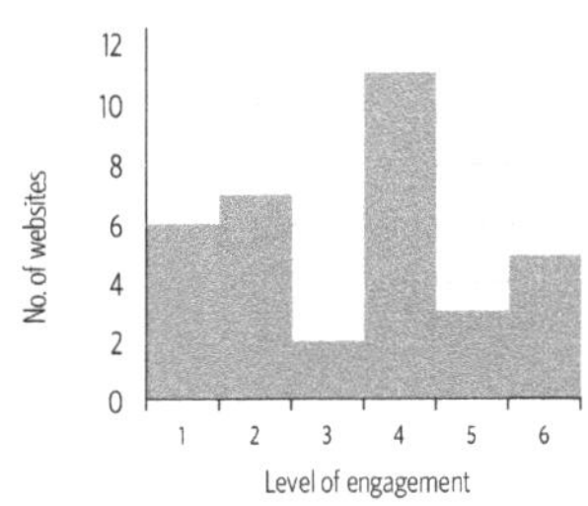

Of the 14 unique websites that published the 34 web pages, 2 stand out as publishing web pages in which the authors of those pages engage closely with the content of an open access journal article: (1) Child Health Safety and (2) Vaccine Papers. Authors of articles published on these two anti-vaccination websites present findings in their own words, they reinterpret findings by engaging critically with the methods and/or analyses presented in the original study, often focusing on a scientific paper that shows no association between vaccination and autism, and they refer to other scientific research to support their reanalysis. An additional mark of their close engagement with the journal article is that the authors of these web pages reply to questions and challenges posed in the comments section of the web page, often posted by pro-vaccination individuals, in order to provide additional clarity in support of their reanalyses.

A comparison between the level of engagement of selected websites (that is, for those that contained more than three web pages in the sample) and the activity of their Twitter accounts shows that those websites that are the most engaged are also those with the lowest levels of activity on Twitter. This may suggest that members of the anti-vaccination movement select different online media as their preferred mode of communication depending on how closely they engage with scientific articles. Put differently, social media platforms such as Twitter may be selected by those in the anti-vaccination movement who do not wish or need to engage closely with the scientific content but who nevertheless seek to leverage science using the affordance of that online communication network to further their cause.

Summary of findings

The findings show that the anti-vaccination movement constitutes a relatively small and variable proportion of the total number of social media accounts mentioning scientific articles on Twitter. For one group of open access journal articles, 11% to 18% of Twitter accounts were found to be anti-vaccination; in the case of a second group of open access journal articles, the proportion of anti-vaccination Twitter accounts was found to be between 1% and 3%. This indicates an interest in selected journal articles.

For those open access journal articles with a higher proportion of mentions from anti-vaccination accounts on Twitter, it was found that the activity (proportion of mentions) for those accounts exceeded their representation (proportion of unique anti-vaccination Twitter accounts). In other words, anti-vaccination Twitter accounts tweet more than their representation suggests. This indicates that the anti-vaccination movement 'punches above its weight' when a scientific article that in all likelihood supports its ideological position is accessible and fed into the flow of information in its social media networks.

Despite high levels of activity on Twitter, the level of engagement by the anti-vaccination movement with open access journal articles is low. The frequent reposting of content and the relatively small proportion of original content is the main contributor to low levels of engagement on Twitter. In addition to frequent reposting, evidence was found of content being modified and blended to reignite levels of attention, if not engagement.

It was found that the low levels of engagement by the anti-vaccination movement with open access journal articles in the social media cannot be generalised to similarly low levels of engagement on the web. Overall, engagement by the anti-vaccination movement tends towards appraisal and application in the case of web pages. Two exceptional cases were found in which the authors of anti-vaccination web pages engage closely with open access journal articles. Findings also suggest that those highly engaged with scientific content on the web are relatively inactive in the social media.

Discussion

The production of uncertainty: Selective use of scientific information

In certain circumstances, being confronted by uncertainty has negligible consequences – decision-making can simply be deferred or suspended. But for the uncertain parent of a new-born child who faces a time-bound decision on whether to vaccinate, being confronted by anti-vaccination messaging seemingly supported by science, presents the very real possibility of the parent electing not to vaccinate. And this decision would be taken despite the fact that the consensus position within science, based on available scientific evidence, is supportive of vaccination:

> The broader public health implications of propagating these memes and articles make anti-vax activities more than a bizarre online curiosity. Most of the material that the [...] accounts tweet are designed to erode confidence in vaccination. The goal is to make new parents question everything ... Public health

> officials are concerned. […] [I]t is essential that when people go online for information they are left with the clear impression that vaccines are safe and effective.' If that's going to change, the people fighting misinformation need to understand how it gets spread in the first place. (DiResta & Lotan, 2015)

The findings on the number of mentions to specific articles indicate that the anti-vaccination movement is not only using scientific content in its online communications, but that it is selective in terms of the scientific content fed into their information flows.

The point to note here is that the information accessed is from journal articles produced early in the science communication process at a stage when truth claims are still contested and in flux. This finding confirms the observation that publics with a limited understanding of how science works – in particular, that science is iterative and self-correcting – may select and exaggerate the findings of individual studies (Kahan et al., 2017) and supports the large body of work in the science communication literature on 'phenomena of selection' (Akin & Landrum, 2017: 455). The anti-vaccination movement, rather than being made to wait for settled truth claims to emerge at the end of the iterative and progressive science communication continuum (Cloître & Shinn, 1985, in Bucchi, 2004), accesses 'unsettled', single-study truth claims, and interpret and share them as universal truth.

These 'scientific truths' hold value for the anti-vaccination movement because they confer legitimacy to its cause in the eyes of other non-scientific communication networks. According to Castells (1996), value is what the network determines it to be. In the case of the global anti-vaccination network, which could be situated within the larger global anti-establishment communication network, only that information that supports the beliefs of the network holds currency and is therefore worth exchanging. In the case of scientific information, the findings of this research show that those scientific articles that articulate a causal relationship between vaccinations and adverse health or that express doubt about the efficacy and safety of vaccines, gain currency in the network. Conversely, scientific articles that disprove the dangers of vaccination hold no value and therefore do not circulate within the communication network of the anti-vaccination movement.

While the exchange of selected scientific information on the effects of vaccination serve to reinforce the belief systems of the anti-vaccination movement, they have the opposite effect on other networked communities who are both present in social media networks and therefore inevitably connected to the online anti-vaccination movement. The selective harvesting of information from open science and the communication of that information in social media networks, produces uncertainty in other online communities, even when there is consensus within the scientific community, as is the case for vaccine safety. The production of uncertainty can therefore be understood as an attack on the information flows of other networks and is aimed at destabilising certainty in the information that circulates in those networks.

The amplification of uncertainty

In the theory on communication, amplification is the process of intensifying or attenuating signals during the transmission of information (Kasperson et al., 1988). The amplification of uncertainty using information from scientific sources takes place in the online communications of the anti-vaccination movement by means of at least two mechanisms that are supported by the findings of this study: (1) high levels of activity in online communication networks, and (2) low levels of engagement with scientific information in those networks.

High levels of activity

In general, the most influential tweeters are more active than the less followed tweeters although it is not clear whether these individuals are widely followed due to their high posting volume, or whether they are prolific because their audience is sufficiently large (or appreciative) (Thelwall et al., 2013).

High levels of activity in the social media also increase the probability of content consistent with a particular stance appearing at the top of the content feeds of those who follow highly active accounts. Being listed at the top of content feeds, in turn, increases the chances of the content being shared with others in the social media network (Lerman & Hogg, 2014).

Kumar et al. (2018) have shown that highly active members in online communities are more likely to initiate interaction and conflict with other communities. However, 'while these interactions are initiated by the highly active users of the source community, the attackers and defenders who actually get mobilized to participate in the negative mobilization are much less active than them' (Kumar et al., 2018: 5). This suggests that those highly active anti-vaccination Twitter accounts are not only more likely to instigate interaction with the online pro-science community, but that they play an important role in mobilising less active members who, by taking up the cause, further proliferate the information flows of the anti-vaccination movement in the broader social media network.

The findings of this research show that the anti-vaccination movement 'punches above its weight' when a scientific article that supports its ideological position is accessible and inserted into its networked information flows on Twitter. In other words, the proportion of tweets by the anti-vaccination movement for selected open access journal articles is higher than the proportion of anti-vaccination Twitter accounts that mention the same open access journal article. It is the structure and affordances of networked communication that makes possible such 'network effects'.

The distribution of activity for the three most-mentioned articles show a skewed distribution in which most Twitter accounts mention an article only once. However, the data also show that for all three of these open access journal articles, there are a few Twitter accounts that mention the article more than 10 times, and in all three cases there is one Twitter account that mentions the article 100 or more times.

The most active Twitter account in the anti-vaccination movement was found to share the same content repeatedly across the network for an extended period of time. And, in general, most content consists of retweets and reposts. If we take as a starting point that the anti-vaccination movement in terms of its network properties is highly homogeneous, then research by Piedrahita et al. (2018) on how contagion dynamics emerge when networked actors repeatedly contribute to activity around a collective cause may be significant. They conclude that 'to the extent that digital technologies are inserting networks in every aspect of social life, our results suggest that we should expect to see more instances of large-scale coordination cascading from the bottom-up' (Piedrahita et al., 2018: 334). And according to Asur et al. (2011), rather than a large number of followers, the most effective strategy to propagate information (at least in terms of creating trending topics) on Twitter is to retweet; the number of retweets for a topic correlates strongly with the length of time the attention of the network is held.

Low levels of engagement

The finding of low levels of engagement on Twitter, this suggests that scientific content is treated at face value, and that scientific information flows through social media networks with little need for actors in these communication networks to engage deeply with the information presented to them. This finding confirms findings by Thelwall et al. (2013) that the content of tweets linking to journal articles are unlikely to contain insightful responses to the content of those articles.

The explanation for face-value engagement, according to Stalder (2006), describing the work of Castells, is that certain networks rely and depend on information being taken at face value. Trusted intermediaries are established as central nodes in the communication network; they facilitate the rapid exchange and transfer of information that can be trusted at face value across the network.

Different programs of networks determine the speed at which information is accepted as accurate before it is acted upon. The global science network, for example, is programmatically sceptical – scientists are more likely to interrogate information received from others in the network before acting. Social media networks are less so because they are programmatically a network of attention (Wu, 2016), and to retain attention, information must flow constantly regardless of the accuracy of the information. Neither ideologically-motivated social movements nor the owners of social media platforms, both of whom are locked into attention-seeking behaviour, derive value by adhering to the norms accepted within science as being necessary for establishing the truth.

The findings of this research support the proposition that scientific information is taken at face value in online communication networks. There is little evidence of engagement, enabled by affordances such as retweeting, and this allows information to flow at high speed between and across communication networks by enabling high levels of online communication activity.

In sum, a social movement that holds a view contrary to that of science (in this case, the anti-vaccination movement) is both highly active and selective in terms of the information accessed from open science and fed into its communications in the social media. The movement produces uncertainty in online communication networks; uncertainty that cascades across online communication networks programmed for attention and devoid of the normative guidance of science institutionalised for settling truth claims.

Risks and implications for science communication

Risk can be amplified by social factors when risk signals are received, interpreted and passed on by a variety of social actors (Kasperson, et al., 1988). Previous research has interpreted social amplification effects as being place-bound (Petts & Niemeyer, 2004). However, if social amplification is dislocated from place and takes place in the space of flows exemplified by online social (media) networks, then the amplification of risk may be increased.

The amplification of risk in the social media (space of flows) has a bearing on how change is effected in the real world (space of places). Miller (2017) argues that social media communication is not transformative but phatic and, as a consequence, the social media do not mobilise political action. The assumption is that change proceeds from the real-time, global interpersonal connections and communications in the space of flows made possible by the social media to action in the space of places against concentrated, hierarchically structured power such as an oppressive regime or Wall Street. In other words, change requires a switch from the space of flows to the space of places. Based on a review of the evidence, Miller (2017) argues that social media activism is not transformative or politically goal-orientated.

The evidence presented in this chapter suggests that the social media can effect change along different lines. Change is made possible by the production of uncertainty and, in certain spheres of social life such as health and well-being, is equally capable of driving change. Change proceeds from the real-time, global interpersonal connections and communications not to place-bound action but to an increase in influence over a diffused and interconnected mass public, more of whom seek meaning in the space of flows (Castells, 2009; Stalder, 2006). In this sense, every connected individual is a target in attacks on what is held to be certain. In cases where such attacks generate uncertainty, the potential arises to alter the decisions taken by individuals. In this change process, a shift from the space of flows to space of places is not required because change is effected by disrupting the flows of information in other communication networks through the creation of uncertainty. This includes disruption in the flows of scientific certainty.

It is unlikely that all change can be effected in this manner but it is short-sighted to suggest that the social media cannot be used by social movements to effect change, particularly when decision-making at the individual level poses risk to entire populations.

Unlike the global financial network that has created centrally located and trusted nodes in the networked flows of information to ensure that information in the network can be taken at face value, as well as self-regulating structures within the network and the assurance of external, government intervention in the case of the threat of collapse, social media networks have no such mechanisms to ensure trust in the information exchanged.[15] This may account for why ideologically-motivated social movements active in global social media networks appear to borrow or import trust from the open communication networks of a social institution that is trusted by the public: science.[16] As the Vaccine Research Library proclaims: 'We have more than 7,000 links to abstracts and full text from mainstream, scientific literature [...]. If their own literature isn't a 'reliable source', then what is?' (Vaccine Research Library, 2015: n.p.).

In the real world (the space of places), social cues confer authority and trust; in networks (the space of flows), these cues are not necessarily linked to class, cultural status or other traditional social cues (Lin, 2008). Network social capital or 'network capital' may present itself as a new type of capital to emerge alongside other types of social capital (Bourdieu, 1986) that accumulates in virtual networks as socially networked actors attract and consolidate the attention of others in the network.

Attention-seeking as a strategy to gain an influential position in networks accounts for the migration of unexpected actors to parts of the communication network where they are most likely to attract attention. Medical professionals and scientists are not immune to such attention-seeking behaviour in their quest to extend their influence over others, and it is for this reason that there are doctors and scientists to be counted in the online communication networks of the anti-vaccination movement. In some instances, existing capitals (cultural or symbolic) from the space of places may be leveraged to attract attention in the

15 It is not that the structures put in place by the global financial network are infallible. The point is that there have been attempts to self-correct, that is, to protect the network's program of surplus accumulation. There has been no concerted attempt to self-correct across social media networks.

16 See Schäfer (2017) on science as a trusted institution and Lin (2008) on open networks and the accumulation of social capital.

space of flows (networks) so as to accumulate network capital. For example, doctors converting their cultural capital (expertise) to network capital by switching allegiance from the professional network of medicine to that of the anti-vaccination movement's communication network.

There is also the potential for the conversion of network capital to economic capital. Fake photo sites on Twitter post doctored historical photos. The photographs posted are known to be fake but nevertheless hold popular appeal. Once these Twitter accounts have amassed a large number of followers, they leverage their network capital to attract economic capital as advertisers are prepared to provide financial rewards for the attention that these accounts can bring to their brands via their follower networks.[17] In other words, social media does not conform to expected rules and social hierarchies that confer authority or trust – a fake account can attract more attention and, by implication, yield more influence in a social media network than the account of a trusted, authoritative source, including that of a scientist.

The motivations behind scientists and medical professionals' participation in the social media or their motivations for expressing their allegiance to the anti-vaccination movement, may well be attention-seeking. Nevertheless, their presence and their inferred authority in these communication networks, destabilises traditional social cues of authority by creating the perception of divided positions on which there is, in reality, scientific consensus. In such a scenario, who to trust becomes unclear. It may be clearer to establish trust within relatively closed networks with shared norms and values (Burt, 2001), but to outsiders, where those norms are no longer shared, it becomes increasingly difficult to identify trusted sources. This has implications for uncertain parents and policy-makers alike who find themselves participating in online communication networks where trust has been destabilised, as this research has shown, by active minority

17 See Reply All podcast #48 by Gimlet: https://gimletmedia.com/episode/48-i-love-you-i-loathe-you/

groups exploiting the attention imperatives of those networks. As Southwell (2017: 223) states, communication of scientific information in the social media 'can undermine scientific authority, complicate decision-making and fuel the propagation of rumours and misinformation'.

At stake is the credibility of science as an institution in the eyes of the public (Kahan et al., 2017). Institutions such as science react to threats to their credibility (the extent to which they are trusted) by making taken-for-granted norms more explicit (Weingart, 2017). In network terms, the network's programmers must defend the logic of the network from attacks made by other networks or emanating from the network environment (Stalder, 2006). By making the institutional norms of science more explicit or by enforcing the terms of participation in the global science communication network, scientists should, in theory, refrain from non-normative attention-seeking in other communication networks or face sanction from their own.

How centrality is established in networks and how 'network capital' is accumulated to establish a position of trust and influence in an online social network, a question that was not explored in this chapter, remains opaque. If, as some have suggested (Muller, 2017), influence is a new form of power in the network society, then it becomes increasingly important to understand better not only who the trusted influencers are, but how they establish and protect their positions of influence.

This research has shown how the anti-vaccination movement is able to attract disproportionate levels of attention in online communication networks to exert influence over what is certain or true. Further research and conceptual development are needed to move towards a more comprehensive theory of attention, influence and power in the network society. Developing such an understanding will be critical for the science of science communication as it seeks to inform effective strategies for the communication of science to networked publics.

Conclusion

The research presented in this chapter was a first attempt at creating a better, empirically-based understanding of new potentials in a changed science communication environment; specifically, the potentials arising from increased access by non-scientists to the formal communication of science.

The evidence presented points to the use of selected scientific information, extracted with little engagement from open access journal articles by a highly active minority group to produce and amplify uncertainty in the broader population using social media networks. The social media environment, devoid of scientific norms to steer action toward the establishment of truth, provides an ideal communication substrate, as does the networked nature on online communications. Online communication networks in the form of the social media enable relatively small social movements to exploit the affordances of those networks to amplify their messaging.

That the research focused on a non-scientific social movement opposed to vaccinations meant that the potentials identified were in the form of risks. The study of other social movements' use of the products of open science may reveal more positive potentials. Similarly, research on other social movements may confirm the findings of the single case presented in this chapter. Both endeavours are needed to be able to assert more generalisable insights to advance the science of science communication and to design effective strategies for the communication of science in society shaped by communication networks.

Acknowledgements

This work is based on the research supported by the South African Research Chairs Initiative of the Department of Science and Technology and National Research Foundation of South Africa (grant number 93097). Any opinion, finding and conclusion or recommendation expressed in this material is that of the author and the NRF does not accept any liability in this regard.

References

Akin, H. & Landrum, A. (2017). A recap: Heuristics, biases, values, and other challenges to communicating science. In K. Jamieson, D. Kahan & D. Scheufele (eds), *The Oxford Handbook of the Science of Science Communication* (pp. 455–460). New York: Oxford University Press.

Amer, K. & Noujaim, J. (Directors) (2019). *The Great Hack* [documentary]. USA: Netflix.

Asur, S., Huberman, B., Szabo, G. & Wang, C. (2011). Trends in social media: Persistence and decay. *arXiv*, 1–12. https://arxiv.org/abs/1102.1402

Bean, S. (2011). Emerging and continuing trends in vaccine opposition website content. *Vaccine, 29*, 1874–1880.

Bennato, D. (2017). The shift from public science communication to public relations: The Vaxxed case. *Journal of Science Communication,* 16(2), 1–11.

Blöbaum, B. (2016). Key factors in the process of trust: On the analysis of trust under digital conditions. In B. Blöbaum, *Trust and Communication in a Digitized World: Models and concepts of trust research* (pp. 3–26). Heidelberg: Springer.

Bourdieu, P. (1986). The forms of capital. In J. Richardson (ed.), *Handbook of Theory and Research for the Sociology of Education* (pp. 241⊠258). New York: Greenwood.

Brossard, D. (2013). New media landscapes and the science information consumer. *PNAS,* 10(3), 14096–14101. doi:10.1073/pnas.1212744110.

Bucchi, M. (2004). Communicating science. In M. Bucchi, *Science in Society: An introduction to social studies of science* (pp. 105–123). London: Routledge.

Bucchi, M. (2018). Credibility, expertise and challenges of science communication 2.0. *Public Understanding of Science,* 26(8), 890–893. doi:10.1177/0963662517733368

Burt, R. (2001). Structural holes versus network closure as social capital. In N. Lin, K. Cook & R. Burt, *Social Capital: Theory and research* (pp. 31–56). New York: Aldine De Gruyter.

Cameron, J. (2018, May 22). Controversial short seller Viceroy targets Capitec AGAIN! Read explosive letter to its auditors here. *BizNews.* https://www.biznews.com/sa-investing/2018/05/22/short-seller-viceroy-nails-capitec-again

Castells, M. (1996). *The Rise of the Network Society.* Oxford: Blackwell.

Castells, M. (2007). Communication, power and counter-power in the network society. *International Journal of Communication,* 1, 238–266. http://ijoc.org/index.php/ijoc/article/viewFile/46/35

Castells, M. (2009). *Communication Power.* Oxford: Blackwell.

Chan, M., Jones, C. & Albarracin, D. (2017). Countering false beliefs: An analysis of the evidence and recommendations of best practices for the retracting and correction of scientific information. In K. Jamieson, D. Kahan & D. Scheufele (eds), *The Oxford Handbook of the Science of Science Communication* (pp. 341–349). New York: Oxford University Press.

Dickel, S. & Franzen, M. (2016). The 'problem of extension' revisited: New modes of digital participation in science. *Journal of Science Communication,* 15(1), 1–15. doi:10.22323/2.15010206.

DiResta, R. & Lotan, G. (2015, June 8). Anti-vaxxers are using Twitter to manipulate a vaccine bill. *Wired.* https://www.wired.com/2015/06/antivaxxers-influencing-legislation/

Edelman Trust Barometer (2019) *Edelman Trust Barometer Global Report.* https://www.edelman.com/sites/g/files/aatuss191/files/2019-03/2019_Edelman_Trust_Barometer_Global_Report.pdf?utm_source=website&utm_medium=global_report&utm_campaign=downloads

Fecher, B. & Friesike, S. (2013, May). Open science: Five schools of thought. RatSWD Working Paper Series, pp. 1–11.

Friesike, S., Widenmayer, B., Gassmann, B. & Schildhauer, T. (2015). Opening science: Towards an agenda of open science in academia and industry. *Journal of Technology Transfer* 40, 581⊠601. doi:10.1007/s10961⊠014⊠9375⊠6.

Haustein, S. & Costas, R. (2015). Identifying Twitter audiences: Who is tweeting about scientific papers? *ASIS&T SIG/MET Metrics 2015 Workshop*, (pp. 1–3).

Haustein, S., Bowman, T. & Costas, R. (2016). Interpreting 'altmetrics': Viewing acts on social media through the lens of citation and social theories. In C. Sugimoto (ed.), *Theories of Infometrics and Scholarly Communication* (pp. 372–406). Berlin: De Gruyter. doi:10.1515/9783110308464-022.

Illing, S. (2018, April 4). Cambridge Analytica, the shady data firm that might be a key Trump-Russia link, explained. *Vox.* https://www.vox.com/policy-and-politics/2017/10/16/15657512/cambridge-analytica-facebook-alexander-nix-christopher-wylie

Jasanoff, S. (2006). Transparency in public science: Purposes, reasons, limits. *Law and Contemporary Problems,* 69(21), 21–45.

Kahan, D., Scheufele, D. & Jamieson, K. (2017). Introduction: Why science communication? In K. Jamieson, D. Kahan & D. Scheufele (eds), *The Oxford Handbook of The Science of Science Communication* (pp. 1–11). New York: Oxford University Press.

Kata, A. (2012). Anti-vaccine activists, Web 2.0, and the postmodern paradigm: An overview of the tactics and tropes used online by the anti-vaccination movement. *Vaccine,* 30, 3778–3789.

Kumar, S., Hamilton, W., Leskovec, J. & Jurafsky, D. (2018). Community interaction and conflict on the web. *WWW 2018: The 2018 Web Conference, April 23–27, 2018, Lyon, France* (pp. 1–11). New York: ACM.

Leask, J. (2015). Should we do battle with anti-vaccination activists? *Public Health Research & Practice,* 25(2), e2521515. doi:10.17061/phrp2521515.

Leonelli, S., Spichtinger, D. & Prainsack, B. (2015). Sticks and carrots: Encouraging open science at its source. *Geography and Environment,* 2, 12–16. doi:10.1002/geo2.2.

Lin, N. (2008). A network theory of social capital. In D. Castiglione, J. Van Deth & G. Wolleb (eds), *The Handbook of Social Capital* (pp. 50–69). Oxford: Oxford University Press.
Merton, R. K. (1968). *Social Theory and Social Structure.* New York: Free Press.
Miller, V. (2017). Phatic culture and the status quo: Reconsidering the purpose of social media activism. *Convergence,* 23(3), 251–269. doi:10.1177/1354856515592512.
Moran, B., Lucas, M., Everhart, K. & Morgan, A. (2016). What makes anti‐vaccine websites persuasive? A content analysis of techniques used by antivaccine vaccine sentiment. *Journal of Communication in Healthcare*, 9(3), 151–163. doi:10.1080/17538068.2016.1235531.
Muller, J. (2017). Universities and the 'new society'. In J. Muller, N. Cloete & F. van Schalkwyk (eds), *Castells in Africa: Universities and Development* (pp. 17–31). Cape Town: African Minds.
Ortiz-Ospina, E. & Roser, M. (2016). Trust. *Our World in Data.* https://ourworldindata.org/trust
Popper, K. (1962). On the sources of knowledge and of ignorance. In *Conjectures and refutations*, 3–40. Oxford: Oxford University Press.
Power, M. (1997). *The Audit Society.* Oxford: Oxford University Press.
Power, M. (2000). The audit society – second thoughts. *International Journal of Accounting,* 4, 111–119.
Schäfer, M. (2017). How changing media structures are affecting science news coverage. In K. Jamieson, D. Kahan & D. Scheufele (eds), *The Oxford Handbook of the Science of Science Communication* (pp. 51–59). New York: Oxford University Press.
Schäfer, M., Füchslin, T., Metag, J., Kristiansen, S. & Rauchfleisch, A. (2018). The different audiences of science communication: A segmentation analysis of the Swiss population's perceptions of science and their information and media use patterns. *Public Understanding of Science*, 1–21. doi:10.1177/0963662517752886.
Scheufele, D. (2013). Communicating science in social settings. *PNAS,* 10(3), 14040–16047.
Scheufele, D. (2014). Science communication as political communication. *PNAS,* 111(4), 13585–13592.
Southwell, B. (2017). Promoting popular understanding of science and health through social networks. In K. Jamieson, D. Kahan & D. Scheufele (eds), *The Oxford Handbook of the Science of Science Communication* (pp. 223–231). New York: Oxford University Press.
Stalder, F. (2006). *Manuel Castells.* Cambridge: Polity.
Streitfeld, D. (2017, May 20). 'The internet is broken': @ev is trying to solve it. *The New York Times.* https://www.nytimes.com/2017/05/20/technology/evan‐williams‐mediumtwitter‐internet.html
Taubert, N. & Weingart, P. (2017). Changes in scientific publishing. In P. Weingart & N. Taubert (eds), *The Future of Scientific Publishing* (pp. 1–31). Cape Town: African Minds.
Tharoor, I. (2018, March 22). The scary truth that Cambridge Analytica understands. *The Washington Post.* https://www.washingtonpost.com/news/worldviews/

wp/2018/03/22/the-scary-truth-that-cambridge-analytica-understands/
Thelwall, M., Tsou, A., Weingart, S., Holmberg, K. & Haustein, S. (2013). Tweeting links to academic articles. *Cybermetrics: International Journal of Scientometrics, Informetrics and Bibliometrics,* 17, 1–8.
Van Schalkwyk, F. (2019). New potentials in the communication of open science with non☒scientific publics: The case of the anti☒vaccination movement Unpublished doctoral thesis, Stellenbosch University.
Weingart, P. (2011). Science, the public and the media – Views from everywhere. In M. Carrier & A. Nordmann (eds), *Science in the Context of Application* (pp. 337–348). Dordrecht: Springer.
Weingart, P. (2012). The lure of the mass media and its repercussions on science. In S. Rödder, M. Franzen & P. Weingart (eds), *The Sciences' Media Connection: Public communication and its repercussions* (pp. 17–32). Dordrecht: Springer.
Weingart, P. & Guenther, L. (2016). Science communication and the issue of trust. *JCOM,* 15(5), 1–7.
Williams, J. (2018). *Stand Out of My Light: Freedom and resistance in the attention economy.* Cambridge: Cambridge University Press.
Winowatan, M., Zahuranec, A.J., Young, A. & Verhulst, S. (2019). Index: Trust in Institutions 2019. The Living Library. GovLab. https://thelivinglib.org/index-trust-in-institutions-2019/
World Health Organization (2019). Ten threats to global health in 2019. World Health Organization. https://www.who.int/emergencies/ten-threats-to-global-health-in-2019
Wu, T. (2016). *The Attention Merchants: The epic struggle to get inside our heads.* London: Atlantic.
Zimmerman, R., Wolfe, R. & Fox, D. (2005). Vaccine criticism on the world wide web. *Journal of Medical Internet Research, 7*(2). https://www.ncbi.nlm.nih.gov/pmc/articles/PMC1550643/

9 Why impact evaluation matters in science communication: Or, advancing the science of science communication

Eric Allen Jensen

Introduction

Science communication activities have different agendas, audiences and venues, but most share the goal of making scientific or technical knowledge and research more accessible for public audiences to understand, discuss or debate. But this leaves open the fundamental questions: What counts as effective science communication? What difference is our science communication making? How can we measure whether the communication approach was effective at developing impact? These questions are fundamental to the science communication enterprise (see, for example, National Science Board [2006]), as their answers provide the pathway to improvement in practice over time.

Impact is the overall net outcomes or results of an activity or intervention (intended or unintended) for individuals or groups. Note that changes or 'impacts' can be in negative or dysfunctional directions (Jensen, 2015b). Impacts could include, for example, development in learning about a specific topic; attitude change; a greater sense of self-efficacy; enhanced curiosity or interest in a subject; and improved skills or confidence.

Despite over two centuries of public science communication practice, there is no consensus on what counts as 'success' for public engagement and informal science learning initiatives.[1] The lack of good evaluation practice across the field is certainly a key contributor to this state of affairs (Jensen, 2015a). Industry standard evaluation conducted across the science communication field comprise a rogue's gallery of errors and poor practice at each stage in the process from research design to instrument design, sampling, analysis and interpretation. This problem of failing to employ established best practice within social research methods to the challenge of evaluating science communication outcomes for audiences extends to other related fields such as museums. When Davies and Heath (2013: 13) reviewed summative evaluation reports produced by numerous UK museums and consultants with the hope of finding 'golden nuggets' of insight, they instead concluded that evaluation 'evidence used to suggest learning or particular forms of learning can appear fragile at best'.

Indeed, low-quality evaluation evidence, as well as the absence of evaluation or evidence-based design of science communication initiatives has been setting up the global enterprise of science communication for failure over many years. Science communication practitioners are rarely trained to be able to distinguish good from bad evaluation methods, and science communication institutions (including funders) are generally uncritical consumers (and producers) of evaluation research. Generally they simply accept results that align with what they wish to believe, without looking too deeply at the methodological rigour underpinning the knowledge claims.

Of course, measuring the impact of science communication on self-efficacy, learning, attitudes and other outcome variables can be challenging. Measuring such impact often requires direct measurement of visitors' thinking or attitudes before and after the science communication activity. However, this direct pre- and post-impact evaluation is rarely implemented within

1 Science communication may be traced back to the Royal Institution Faraday: kingsciencepublic/2018/10/22/triangulating-the-history-of-science-communication-faraday-marcet-and-smart/

science communication practice. Instead, biased feedback survey questions prompting skewed positive responses from audience members comprise the vast majority of evaluation efforts in science communication globally. Poor survey design is routinely used by consultants and practitioners with decades of experience working for top science communication organisations and with funding from pre-eminent scientific institutions and funders. So why don't these institutions and funders demand better? Why don't they apply the same rigorous expectations of scientific research to the communication activities conducted by and on behalf of the same scientific institutions?

Many excuses have been proposed for the widespread lack of methodological quality in science communication evaluation (Jensen, 2015b). However, it is clear that evaluation and social research methods more broadly have not been prioritised in the training of science communicators, despite the centrality of evaluation to good evidence-based practice. Good science communication requires planning that is rooted in the existing knowledge base for science communication, including both theory and research (Dam et al., 2015). It also requires clear objectives from practitioners at the outset in order to establish communication methods that are logically aligned to the aims. Moreover, evaluation results must be planned into the process in order to inform science communication practice. This kind of evidence-based science communication holds real potential for advancing the field over time, if science communication training and education can be enhanced to enable it.

Conducting effective evaluations that accurately measure the intended outcomes and inform practice requires training in relevant aspects of social scientific research methods such as survey design and qualitative data analysis (Krippendorff, 2013). At a more basic level, however, good science communication evaluation requires clear, realistic objectives as a starting point to designing effective measurement tools.

Developing good, evidence-based science communication from the ground up

In order to evaluate a science communication intervention's effectiveness, one first needs to specify the desired outcomes (short and long term). There is a broad range of implicit aims underpinning public engagement with science. However, these aims are rarely made fully transparent to audiences or even to those involved in conducting the activities. This lack of clarity about aims and the logical connections to the science communication activity at hand is widespread across different types of science communication practice (Jensen & Holliman, 2016; Kennedy et al., 2018).

Science communicators should have the end goal in mind, even if that is distant from the initiative itself. Clear definition of aims at a practical level is essential to establish the foundations for effective public communication practice and evaluation (Holliman, 2017b).

The value of taking a systematic approach to defining the nature and level of outcomes for a given public engagement activity includes the following:

- Enables *assessment of success.* Having transparent goals and aims helps to focus the engagement practice itself and to measure the level of its success. This includes checking whether the activity is reaching intended the types of audiences (Jensen et al., 2015).
- *Improves* engagement practice. Use incoming evaluation evidence to continuously improve methods of engaging audiences.
- *Know the impact of the activity:* The activity may be damaging the aims of the public engagement. High-quality evaluation linked to clear, measurable intended outcomes can ensure that the activity remains on track.

An over-emphasis on outputs only (i.e. what you have done, rather than how audiences have benefited) is a common problem in science communication. Science communicators often assume that their outputs necessarily lead to the hoped-for impacts, thereby limiting the scope for improvement over time. Indeed, 'the possibility of negative impacts, are routinely neglected within science communication evaluation' (Jensen, 2014: 2). It is important to have concrete, realistic outcomes specified in advance, which the science communicator then seeks to translate effectively into practice for the benefit of audiences.

This kind of planning information is essential for establishing the basis for accountability and quality in public engagement with science. The other related key factor is ensuring that public engagement initiatives are evaluated regularly for quality of experience, and at least occasionally for impact. Ideally, 'on-going evaluation tied to real-time results can enable science communication organizations to develop activities that stand a stronger chance of yielding positive impacts' (Jensen, 2015a: 1).

Of course, even the most 'successful' science communication initiatives based on our definition above could have implications or results that the initial sponsors or science communicators might find distasteful. This openness in outcomes is inherent in communication. Yet, there is clearly a great deal that science communicators can do to improve the quality and value of their activities for its audiences. Explicitly articulating intended outcomes can help to reveal gaps between science engagement practices and the logical steps on the pathway to achieving valuable engagement aims.

Clarifying aims to set up effective science communication evaluation

Limitations in existing science communication evaluation practices are rooted in science engagement practice (Jensen, 2014) and in the aims practitioners set for those practices. The practical question science communicators should be asking on an ongoing basis is: 'How could I improve my science engagement activities?'

To answer this question, clear aims and evaluation are needed, which should feed into practice in a continual manner to establish an evidence-based approach to science communication.

Good science engagement requires upstream planning and clear objectives, and this is even more so for evaluation. Moreover, there should be a commitment to making improvements in the programme or activity based on what the evaluation reveals. It is important to start with the big picture:

- Why are you doing your public engagement event?
- What do you want to achieve with it?
- How will you know if you have been successful?
- Are your goals clear, specific and realistic?

These probing questions can help inform the design of better public engagement activities, while also setting the activity up for good evaluation. Evaluative thinking is oriented towards making improvements, based on good empirical evidence on what is working and why. There are a number of good reasons to evaluate, including:

- To build a better understanding of target publics, (e.g. needs, interests, motivations, language).
- To inform plans and to predict which engagement or learning methods and content will be most effective.
- To know whether the objectives have been achieved (and why or why not).
- To re-design the approach to be more effective in future.

Good impact evaluation is systematic and thorough. It tells one how and why particular aspects of a science communication activity are effective. It does *not* provide a binary 'good'/'bad' or 'successful'/'unsuccessful' result. This is because a 'successful' project can always develop the good aspects of their practice further. Likewise, there will be specific aspects of an 'unsuccessful' project or method that were ineffective (and should be avoided in future projects).

Understanding audiences for science communication

A surprisingly under-developed aspect of science communication evaluation is establishing the nature of the audiences that attend engagement events and activities in order to identify social inclusion gaps and take participant needs into account. To do this effectively, it is important to gather data about participants as they enter the science communication activity. A recent example showing why this is so important comes from the study entitled, 'Preaching to the scientifically converted' by Kennedy et al. (2018). This study addressed the question: 'Are UK science festivals attracting a diverse and broadly representative sample of the public?' It presents findings from evaluation studies conducted in three major UK science festivals. This included pre-visit survey data collected from a science festival in eastern England (n = 592), in southern England (n = 171) and in northern England (n = 1011).

The study showed that in contrast to its aim of widening access to science engagement, most visitors to the science festivals were already highly engaged in cultural and scientific events prior to their science festival attendance. For example, in the northern England science festival, 65% reported already attending other science festivals, events, or activities prior to their visit. In comparison, the 2014 national Public Attitudes to Science (PAS) survey found that 3% of its national UK sample reported attending a science festival.

The study also showed high pre-visit levels of interest in science amongst science festival audiences. In pre-visit responses for both the southern (88%) and eastern (92%) festivals, visitors agreed they were personally interested in science. The study also found that adult attendees at the science festivals were substantially more highly educated than the UK population as a whole, and science festival attendees were more economically advantaged than the general population. This study's audience profiling revealed disparities in access to science engagement, which could reinforce social inequality. Prior to this study, this key information was not available to science festival organisers.

South African evaluation examples

In this section of the chapter, two examples of evaluation from South African science communication practice are presented.

Evaluating the impact of a South African MOOC

This example focuses on a massive open online course (MOOC) led by Prof. Anusuya Chinsamy-Turan and published by the University of Cape Town. This course was on the theme of 'Extinctions'. In order to develop the impact evaluation of this MOOC, the organisers needed to clarify the most important impact objectives.

In this case, those were primarily learning-oriented objectives, key 'take home' points relating to the course theme of extinctions. Once these impact objectives had been clarified, a set of Likert-type items were developed to evaluate the progress towards achieving those outcomes with participants (Table 1). Each of these Likert-type questions asked for a response on a scale from *strongly disagree* to *strongly agree*, with a neutral mid-point and a 'don't know / no opinion' option. Each of these items was repeated before the course, in the middle and after the end.

By repeating these statements at three different points matched to the same individual, the evaluation was able to show the relative progress of different individuals through the course on the defined learning outcomes.

Table 1: Likert-type items developed to evaluate progress towards achieving outcomes

Likert-type item used to evaluate MOOC impact objectives	**Outcome measured**
'Extinctions in the last 100 years are the result of mostly natural processes, not human activity'.	Understanding of key learning point (reversed)
'Human behaviour is negatively affecting ecosystems'.	Understanding of key learning point
'Biodiversity is a valuable resource for humans to use'.	Understanding of key learning point
'The environment is important to me'.	General attitude statement relating to the theme of the course

Likert-type item used to evaluate MOOC impact objectives	Outcome measured
'Understanding past extinctions can be important for understanding the effects of the 6th extinction'.	Understanding of key learning point
'I think the 6th mass extinction is already underway'.	Understanding of key learning point
'I think there is little humans can do to prevent the 6th mass extinction from happening'.	General attitude statement relating to the theme of the course (reversed)
'All life on earth will soon come to an end'.	Understanding of nuanced learning point, that is, that the process of extinctions may lead to the end of humans but not to the end of all life (reversed)

World Biotech Tour in South Africa

The World Biotech Tour (WBT) is an ongoing global programme coordinated by the US-based Association of Science and Technology Centers (ASTC), with the goal to demonstrate the relevance, excitement and wonder of biotechnology. It involves students, teachers, science centre professionals and the general public in hands-on activities and discussions about key issues pertaining to biotechnology. In 2017, the WBT travelled to South Africa, with events in different cities across the country. Three science centres from three different cities participated. Sci-Bono Discovery Centre was the lead on this effort, working with Sci-Enza in Pretoria and Cape Town Science Centre.

Different evaluation approaches were used with the different categories of audience for the WBT initiative, with a set of surveys as the primary evaluation approach. The example in this chapter focuses on the Ambassador Programme. For this programme, each of the three science centres assembled a team of high school students (designated as 'ambassadors') to develop and present a biotechnology topic of their choice. A total of 13 students took part in the programme. They were supported by mentors with links to each of the centres, who provided their expertise to help them with their research and presentations. Both ambassadors and their mentors were surveyed as part of the evaluation, taking into account their experience and views. This example addresses the experience and impact for the ambassadors in three of the stages of the programme (pre, mid, and post-programme).

To evaluate the impact and quality of experience for the Ambassador programme during the WBT in South Africa, the following surveys were designed and administered:

- *Pre-programme survey*: This survey included demographic information, interests and motivations relevant to the programme. It also contained outcome measures that were repeated across each survey to allow for evaluation of change over time (i.e. 'impact'). This included measures of biotechnology and general science knowledge and interests, views about scientists and scientific careers, as well as more general transferable skills about confidence and skills they may develop during the programme.
- *Mid-programme survey*: This survey focused on feedback on the experiences of the programme while it was still ongoing to highlight any concerns/issues that should be addressed by the participating science centres.
- *Post-programme survey*: This survey focused on self-report of programme experiences and retrospective assessments to highlight possible areas of improvements for the programme. Additionally, programme impacts were assessed using items that repeated across each survey to show individual-level impacts.

In all cases, initial results were made available to organisers at each of the participating science centres to allow use of the information with pre-programme and intermediate surveys to allow improvements during the programme.

Most evaluations focus only on quality of experience information. In the case of WBT, such feedback questions were included in the survey design in addition to the repeated measures impact items. The post-programme survey items shown below were designed to assess the value the participant placed on the experience of being an 'ambassador' for this programme.

The 'please explain' follow-up questions shown in Figure 1 only appeared when a negative response was submitted by the respondent. In order to go beyond post-visit quality of experience measurement only, repeated measures (pre- and post-test type) were used to evaluate impact by comparing responses before and after the experience (see Figure 2). Each of the questions was repeated exactly at all three data collection points in order to track change over time (i.e. evidence of impact).

Figure 1

	Strongly disagree	Disagree	Somewhat disagree	Neutral	Somewhat agree	Agree	Not applicable/ no opinion
Overall, the Ambassador's programme was a poor use of my time	O	O	O	O	O	O	O
	Please explain:						
Overall, I found the content of the Ambassador's programme useful	O	O	O	O	O	O	O
	Please explain:						

Figure 2

	Strongly disagree	Disagree	Somewhat disagree	Neutral	Somewhat agree	Agree	Not applicable/ no opinion
Science is irrelevant to my life	O	O	O	O	O	O	O
Biotechnology helps to solve the worlds problems	O	O	O	O	O	O	O
Science is not for me	O	O	O	O	O	O	O
Scientific knowledge is important for my future career	O	O	O	O	O	O	O
Biotechnology is hard to understand	O	O	O	O	O	O	O
If I wanted to, I could be a scientist	O	O	O	O	O	O	O
Science is boring!	O	O	O	O	O	O	O

The ambassadors' *pre-programme* survey results indicate that this participant group understood *biotechnology* as something that can solve problems and generate positive change, and considered it to have potential for helping people address everyday issues.

The *mid-programme* results show positive attitudes towards the WBT scheme as participants in this group mentioned advantages of spending time and sharing ideas with scientists and people from other countries. Moreover, the *mid-programme* results evidenced engagement of ambassadors with the WBT programme.

The *post-programme* evaluation shows positive outcomes with the majority of participants highlighting that scientific knowledge is relevant for advancing their future careers. Also, the survey results show that most participants in this group felt they could be scientists if they wanted to, which indicates successful outcomes for the programme in terms of empowering young people. Furthermore, ambassadors in the *post-programme* evaluation indicated that the programme helped them to develop communication and networking skills.

Figure 3a, presents results showing improvement in ambassadors' opinions about their ability to become scientists, and Figure 3b shows positive impact of the programme on their ratings of the importance of science in their careers. In both cases, there is a sharp positive increase from before to after the WBT programme. As a result, the programme has evidently been successful in developing impact on the ambassadors' empowerment and engagement with science. In comparison, the level of impact on attitudes about the relevance of science to ambassadors' daily lives was much less pronounced (Figure 3c). This indicates that the programme is more effective in boosting scientific self-efficacy (the belief in one's capacity to engage with science) than demonstrating to ambassadors the relevance of science to their lives.

This example shows the distinction between widespread quality of experience evaluation, on the one hand, and impact evaluation, on the other.

Figure 3: (a) Self-belief in capacity to be a scientist; (b) Importance of science for future career; (c) attitudes about the relevance of science to daily lives

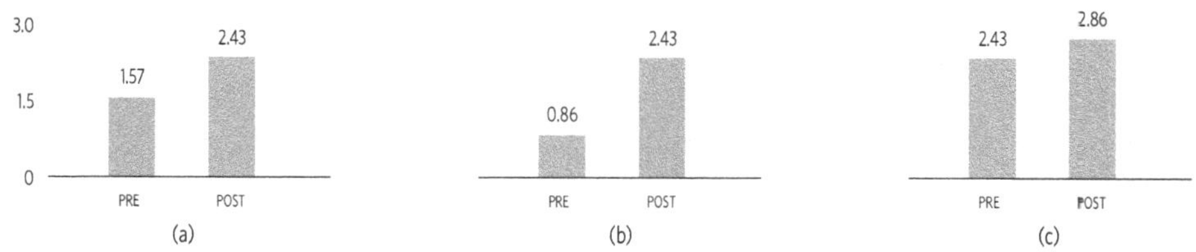

Evaluating science communication impact

Key challenges in evaluating impact apply to both offline and online science engagement:

- Defining the intended impacts for a particular initiative or activity is topic-, audience-, sponsor- and context-specific.
- Clarifying distinctions between exposure, involvement and impact is essential. Often practitioners unintentionally conflate these goals, undermining any effort at impact measurement (e.g. see Jensen, 2015a).
- There may be a gap between explicit and implicit aims, and motivations for the public engagement with science activity (Jensen & Buckley, 2014; Jensen & Holliman, 2016).
- Impacts may be delayed and unfold over time (Jensen et al., 2017).
- Impacts may emerge due to factors after the initial activity/event/content (i.e. what is sometimes called a 'sleeper effect').
- Impacts can be modulated by the socio-economic profile of public engagement participants (Jensen, 2013).
- Measuring long-term impact can be demanding in terms of both expertise and resources (cf. Jensen & Lister, 2015).

It is clear that continuous evaluation practice tied to real-time results can enable science communication organisations to develop activities that are more likely to deliver positive impacts. In light of the barriers science communication organisations face when working to establish high-quality evaluation, current technology linked to good methods can offer a valuable way forward. Given the practical barriers of required expertise, time and resources, continuous impact evaluation can seem like an insurmountable challenge. Yet recent technological improvements have created new means of gathering and analysing ongoing quantitative and qualitative survey-based evaluation using automation (cf. Jensen, 2015a). While social scientific expertise is always required at some points, an automated system can run in an efficient way to provide insights to science communication organisations on an ongoing basis. One example of such an initiative to establish robust technology-enhanced impact evaluation is ZooWise.[2] The ZooWise initiative provides sector-wide, multilingual and widely usable evaluation tools and metrics for zoos, aquariums, botanical gardens, national parks and other nature-oriented public engagement organisations. A similar initiative is ramping up for science communication, called SciWise.[3] This joins other initiatives such as COVES[4] that are aimed at establishing robust methods for sector-wide evaluation.[5]

Conclusion

Developing more effective evidence-based science communication practice will require greater commitment to robust evaluation and making changes to practice on the basis of such evaluation. To begin this process, dramatic improvement in survey design across the international field of science communication is needed (Jensen, 2014). Once good evaluation instruments are established, accurate methods of gathering and analysing data are needed. Throughout this process, it is important to keep in mind that 'success' should not be assumed. 'Given the complexity of science communication interactions – bringing together multiple

2 www.zoowise.org

3 www.sciwise.org

4 www.understandingvisitors.org

5 Practical examples and 'top tips' on evaluation can be accessed at: practicalevaluation.tips

individuals' values, assumptions, world views and meaning-making processes – the remarkable scenario is when positive outcomes are achieved' (Jensen, 2015b: 1). This means that science communication evaluation efforts should start from a neutral standpoint, open to the possibility of both positive and negative outcomes. This standpoint makes it most likely that the evaluation will be useful in highlighting where improvements are needed to make a science communication activity more successful.

The systemic failures in science communication practice must be brought into the light through robust evaluation in order to reveal the pathway to better practices and impacts. At the same time, positive impacts developing from effective practices must be identified systematically in order to develop even more beneficial outcomes. High-quality impact evaluation can be combined with theoretically and empirically informed planning process and ongoing critical self-reflection to enable evidence-based science communication to achieve new heights of positive impact for society (Holliman, 2017a).

References

Dam, F., Bakker, L., Dijkstra, A. & Jensen, E. A. (in press). *Science Communication: A knowledge base*. World Scientific Publishing.

Davies, M. & Heath, C. (2013). *Evaluating Evaluation: Increasing the impact of summative evaluation in museums and galleries*. http://visitors.org.uk/evaluating-evaluation-%E2%80%90-increasing-impact-summative-evaluation-2013/

Davies, M. & Heath, C. (2014) 'Good' organisational reasons for 'ineffectual' research: Evaluating summative evaluation of museums and galleries, *Cultural Trends, 23(*1), 57–69. doi: 10.1080/09548963.2014.862002.

Jensen, E. (2013). Re-considering 'The Love of Art': Evaluating the potential of art museum outreach. *Visitor Studies*, 16(2), 144–159. doi: 10.1080/10645578.2013.827010.

Jensen, E. (2014). The problems with science communication evaluation. *Journal of Science Communication,* 1(2014), C04. http://jcom.sissa.it/archive/13/01/JCOM_1301_2014_C04/JCOM_1301_2014_C04.pdf.

Jensen, E. (2015a). Evaluating impact and quality of experience in the 21st century: Using technology to narrow the gap between science communication research and practice. *JCOM: Journal of Science Communication*, 14(3), C05. https://jcom.sissa.it/archive/14/03/JCOM_1403_2015_C01/JCOM_1403_2015_C05

Jensen, E. (2015b). Highlighting the value of impact evaluation: Enhancing informal science learning and public engagement theory and practice. *JCOM: Journal of Science Communication*, 14(3), Y05. https://jcom.sissa.it/archive/14/03/JCOM_1403_2015_Y05.

Jensen, E. & Buckley, N. (2014). Why people attend science festivals: Interests, motivations and self-reported benefits of public engagement with research. *Public Understanding of Science,* 23(5), 557–573. doi: 10.1177/0963662512458624.

Jensen, E. A. & Lister, T. P. (2015). Evaluating indicator-based methods of 'measuring long-term impacts of a science center on its community'. *Journal of Research in Science Teaching,* 53(1), 60–64. doi: 10.1002/tea.21297.

Jensen, E. & Holliman, R. (2016). Norms and values in UK science engagement practice. *International Journal of Science Education – Part B: Communication and Public Engagement,* 6(1), 68–88. doi: 10.1080/21548455.2014.995743.

Jensen, E. A., Kennedy, E. B. & Verbeke, M. (2015). Outreach: Science festivals preach to the choir. *Nature,* 528(193). [DOI: 10.1038/528193e]

Jensen, E., Moss, A. & Gusset, M. (2017). Quantifying long-term impact of zoo and aquarium visits on biodiversity-related learning outcomes. *Zoo Biology,* 36(4), 294–297. doi: 10.1002/zoo.21372.

Kennedy, E. B., Jensen, E. A. & Verbeke, M. (2018). Preaching to the scientifically converted: Evaluating inclusivity in science festival audiences. *International Journal of Science Education Part B: Communication & Engagement,* 8(1), 14–21. doi: 10.1080/21548455.2017.1371356.

Krippendorff, K. (2013). *Content Analysis: An introduction to its methodology* (3rd edn). Pennsylvania: Sage.

Holliman, R. (2017a). Supporting excellence in engaged research. *Journal of Science Communication*, 16(05), C04.

Holliman, R. (2017b). Assessing excellence in research impact. National Coordinating Centre for Public Engagement Blog. Bristol: NCCPE.

Holliman, R. (2017c). Supporting excellence in engaged research. *Journal of Science Communication,* 16(5). doi 10.22323/2.16050304.

National Science Board (2006). Science and engineering indicators. Washington, DC: US Government Printing Office.

About the editors and the authors

About the editors

Peter Weingart is professor emeritus of sociology and science policy, University of Bielefeld, Germany, and former director of the Institute for Science and Technology Studies (IWT 1993–2009) as well as of the Center for Interdisciplinary Research (ZiF) (1989–1994) at that university. He is currently the South African Research Chair in Science Communication at Stellenbosch University (since 2015) where he was previously a visiting professor (1995–2015). He was a fellow of the Wissenschaftskolleg Berlin (1983/84), a visiting fellow at Harvard University (1984/85) and at the Getty Research Institute (2000), and a fellow of STIAS. He is a member of the Berlin-Brandenburg Academy of Sciences as well as of the Academy of Engineering Sciences (acatech). He was named Distinguished Affiliate Professor of the Technical University Munich in 2012.

Prof. Weingart was appointed editor of *Minerva* in 2007 and is managing editor of the *Yearbook Sociology of the Sciences*. He has also served on the editorial boards of several international journals. Prof. Weingart has published numerous monographs, edited volumes, book chapters and journal articles. Current research interests are science advice to politics, science–media interrelation, and science communication.

Marina Joubert is a senior science communication researcher at the Centre for Research on Evaluation, Science and Technology (CREST) at Stellenbosch University, South Africa, where she is part of a research team associated with the DST/NRF South African Research Chair in Science Communication. Her research interests focus on scientists' role in public communication of science, online interfaces between science and society and the changing policy environment for public communication of science in Africa. She is also interested in the communication of contested topics in science, in particular the vaccine debate. In addition to her teaching and research duties, she presents an annual online course for science communicators across Africa.

Bankole Falade is a researcher with the South Africa Research Chair in Science Communication, Centre for Research on Evaluation, Science and Technology (CREST) at Stellenbosch University. His teaching and research interests include science and society studies; science communication; health communication; science, religion and public life; text mining; and survey research methodology. He was in his previous career a journalist and served as editor of both *Sunday Punch* and *Sunday Independent* newspapers in Nigeria.

List of authors

Doug S. Butterworth is emeritus professor, Department of Mathematics and Applied Mathematics, University of Cape Town.

George Claassen is extraordinary professor, Department of Journalism, Stellenbosch University.

Anne M. Dijkstra is assistant professor in science communication, University of Twente.

Penelope S. Haworth is manager of communication and governance, South African Institute for Aquatic Biodiversity.

Eric Allen Jensen is consulting on three European Commission-funded projects with a strong emphasis on science communication: RRING, Terrifica and GRRIP.

Janice Limson is professor of biotechnology and the DST-NRF South African Research Chair in Biotechnology Innovation and Engagement, Rhodes University.

Wilfred Lunga is a research fellow, Department of Research Use and Impact Assessment, Human Sciences Research Council.

Shirona Patel is head of communications, University of the Witwatersrand.

Konosoang Sobane is a science and health communication specialist, Department of Research Use and Impact Assessment, Human Sciences Research Council.

François van Schalkwyk is a postdoctoral research fellow, Centre for Research on Evaluation, Science and Technology, Stellenbosch University.

www.ingramcontent.com/pod-product-compliance
Ingram Content Group UK Ltd.
Pitfield, Milton Keynes, MK11 3LW, UK
UKHW050845101026
15624UKWH00018B/606

9 781013 294